STAR
WATCH

I DEDICATE THIS BOOK

to all my friends so instrumental
in my early astronomical career:
Patrick Moore, Jack Bennett,
Tom Geary, Michael Sears,
George Buric, Lewis Hurst and
Mary Fitzgerald

to my parents, Phyllis and Leon,
for their great enthusiasm in
purchasing my first astronomical
telescope

to my wife and star on Earth,
Liz, for all her help with the
manuscript and for unfailing
support, encouragement, and
prayer

STAR WATCH

DAVID BLOCK

A LION BOOK

Copyright © 1988 Lion Publishing

Published by
Lion Publishing plc
Sandy Lane West, Oxford, England
ISBN 0 7459 3024 7 (paperback)
Lion Publishing
850 North Grove Avenue, Elgin, Illinois 60120, USA
ISBN 0 7459 1437 3 (hardback)
Albatross Books Pty Ltd
PO Box 320, Sutherland, NSW 2232, Australia
ISBN 0 7324 0832 6 (paperback)

First edition 1988
Reprinted 1994

Printed and bound in Hong Kong

Library of Congress Cataloging-in-Publication Data
Block, David
    Starwatch
    1. Astronomy—Popular works. 2. Cosmology—
Popular works. I. Title
QB44.2.B57 520 88-12729
ISBN 0-7459-1437-3

A catalogue record for this book is available
from the British Library

CONTENTS

FOREWORD

'What is man, that you think of him?' asked the psalmist as he worshipped God, and the question re-echoes with increasing intensity in our own century. As astronomers have unveiled an ever vaster universe, humankind seems to dwindle into a mere speck, a cosmical nonentity lost on a sea of space and time.

Yet, as America's Walt Whitman proclaimed, 'I believe a leaf of grass is no less than the journey-work of the stars.' And today biologists admit that the intricate molecular chemistry of every living thing makes stars look simple in their construction. The human brain is the most marvelously complex assemblage known anywhere in the universe. In some profound sense it seems that humankind was created for the purpose of understanding the cosmos. Indeed, the psalmist goes on to declare, 'Yet you made him inferior only to yourself; you crowned him with glory and honour.'

David Block's book celebrates the paradox of our place in the universe. Following the exclamation that 'the heavens declare the glory of God', he places before the reader some of the most awesome as well as some of the most beautiful images that modern astronomical instruments have revealed. From the nebular wombs of new stars to the filamentary remains of dead and exploded stars, from the strange new worlds of the Voyager spacecraft to the dazzling pinwheel spirals of the distant galaxies, he exhibits the glories of the larger Creation. At the same time he pauses to reflect on the astonishing details that have made the universe a hospitable place for intelligent life and enquiring minds.

The psalmist's question and answer reminds us not only of the immensity of the cosmos, but it affirms our place in the universe as intelligent beings created in the image of God. I certainly hope that as sentient, moral creatures we have enough intelligence and conscience so that future generations can continue to enjoy starwatching and to explore the miracle of creation.

Owen Gingerich
Harvard University and the Smithsonian Astrophysical Observatory

'The Lord made the stars,
    the Pleiades and Orion.
He turns darkness into daylight,
    and day into night.
He calls for the waters of the sea
    and pours them out on the earth.
His name is the Lord.'

From the book of Amos

STARWATCHING

On a clear night away from the glare of city lights, we look up at the starry sky and wonder:

We wonder at the beauty of the heavens;
We are overawed by the immensity of the universe;
We are humbled as we consider our own place as tiny human specks on Earth, a lesser planet towards the edge of one of the millions of galaxies.

Since my school days I have been fascinated with astronomy. I have been fortunate in being able to turn a hobby into a profession, and this book is a collection of some of the marvellous sights and facts I have discovered in my work. I want to share my enthusiasm for starwatching! We will begin by focusing our telescopes on our near neighbours in space, the planets, and then move out into the cosmos, looking at the formation and shapes of stars and galaxies, right to the very edge of the universe.

Yet however majestic the night sky, for some it reflects back a worrying light on their own lives. How can they be important? Looking out on this vast expanse they feel very small and alone. Here again I have been fortunate. The ancient Jewish writers knew that the God who made the heavens also made humanity. We are not insignificant because we are created for a purpose. I have also come to share this faith, and for me the universe is filled with hope as well as with beauty.

But let us begin our journey, both in time and space. And we start close to home, within our own solar system.

Reflected Beauty

PHOTOGRAPHS TAKEN FROM THE
VIKING AND VOYAGER SPACE PROBES
HAVE EXTENDED OUR KNOWLEDGE OF
OUR NEIGHBOURS, THE PLANETS.

The major difference between a star and a planet is that a planet reflects light, whereas stars radiate light. Stars have their own internal power houses which give off light all the time. Planets, on the other hand, do not radiate light, but reflect the incident sunlight falling upon them.

For the purposes of illustration in this book, I have chosen to discuss in some detail three of the planets: Mars, Jupiter and Saturn.

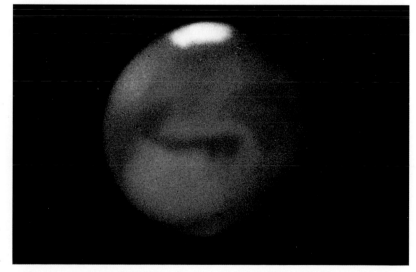

For a long time, scientists have wondered if there is life on Mars.

Of all the planets, Mercury is closest to the blazing surface of the Sun, while distant Pluto is one hundred times further out. The distances from the Sun are shown for each planet and the figures are given in millions of kilometres.

Mercury   58

Venus   108

Earth   149.5

Mars   228

Jupiter   778

Saturn   1427

Uranus   2870

Neptune   4497

Pluto   5900

Of all the planets in the solar system, Mars has aroused the most intense interest as a possible site for extra-terrestrial life. Mars is a small planet, with a diameter just over one-half that of the Earth's. But it closely resembles the Earth in several respects:

- there are the Martian polar caps, which bear a superficial resemblance to the Arctic and Antarctic polar caps on Earth.

- a Martian solar day is only thirty-seven and a half minutes longer than a solar day on Earth.

- the axis about which Mars rotates is tilted by about 24 degrees to the plane of its orbit – much the same as the 23.5 degree axis tilt of the Earth.

It is the tilt of the axis which is responsible for seasons, so that Mars goes through a year with four seasons, just like the Earth, although the Martian year is considerably longer. The distance of Mars from the Sun is 227 million km/142 million miles, and the planet takes just under two Earth years to undergo one complete revolution.

It was the Italian astronomer Giovanni Schiaparelli who, in 1877, published the results of his telescopic obser-vations of Mars. He reported seeing long, faint, straight lines which he called 'canali' or 'channels'. When this Italian word 'canali' was mistranslated into English as 'canals', endless debates and widespread speculation fol-lowed. Were these 'canali' artificial waterways built by an advanced race of Martians?

In 1976, two United States spacecraft, Viking 1 and 2, reached Mars. Both contained an orbiter (which would

The lander of Viking 1 touched down on a rocky plain called Chryse Planitia (the Golden Plains) on 20 July 1976. Forty-five days later, on 3 September, Viking 2 landed on Utopia Planitia (the Utopian Plains).

Sunrise at Utopia Planitia, as transmitted back to Earth by the Viking 2 Lander.

remain in orbit around the planet), and a lander. The Viking landers housed some of the most advanced technology of our day, and were true marvels of miniaturization. They were designed to 'see', 'feel' and 'smell' the Martian environment.

The search for possible extra-terrestrial living organisms in our solar system had now reached its climax. Scrutiny of thousands of photographs from the Viking and earlier probes had failed to reveal any objects or changes which could be attributed to some sort of biological process. But the Viking landers had more than cameras on board. Both landers could manipulate a scoop at the end of a mechanical arm to secure rock and soil samples. In a number of biological experiments, soil was placed in a closed container with or without some nutrient substance, and the container was examined for any changes in its contents.

Living entities alter their environment by eating, breathing, and by giving off waste products. The scientists hoped to detect any minute changes in the make-up of the soil and the experiments were so sensitive that had one of the probes landed anywhere on the Earth (except possibly in the Antarctic), they would easily have detected life. But on Mars no conclusive evidence for life was found. All indications from Viking are that Mars is a sterile and lifeless world.

But what the Viking probes did find was a planet rich in geological detail with majestic volcanoes and huge chasms. The great canyon Valles Marineris is comparable in size to the longest geological fault on Earth, the Rift Valley in Africa. And the largest Martian volcano,

Olympus Mons, rises to an amazing height of 25 km/16 miles above the surrounding plains and has a base of 600 km/375 miles! Yet it is not all volcanoes and canyons. Other areas of Mars reveal dry river beds and evidence of 'flash-flooding'. Running water was once present on the planet.

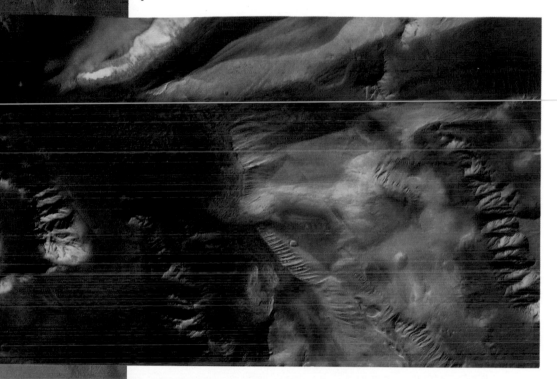

A high-resolution view of the Candor Chasma region of the Martian canyonlands. These canyons were not cut by running water, but are tectonic in origin. They represent splitting of the Martian crust rather than erosion.

Valles Marineris stretches 4,000 km/2,500 miles down towards the towering volcanoes of the Tharsis region. White clouds can be seen filling the 120 km/75 mile wide caldera of the volcano Arsia Mons, the rightmost of the three massive volcanoes stretching across the bottom of the photograph.

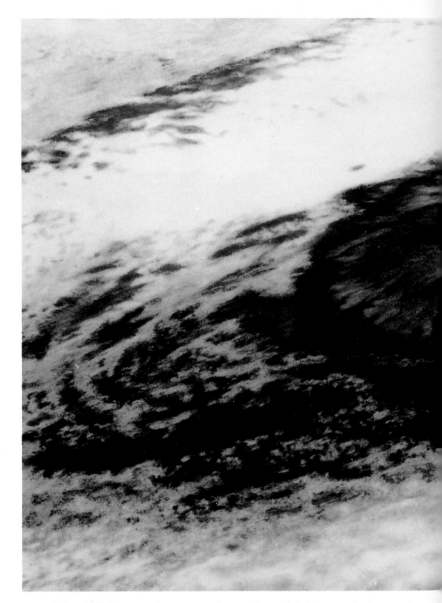

The Viking probes may have found Mars to be a most fascinating planet, but it is a lifeless world, totally different from that once envisaged by H.G. Wells in his novel *War of the Worlds*. In this book Wells depicted the invasion of Earth by aliens from Mars. But there are no creatures on Mars, not even the simplest microbes. The canals are an optical illusion: 'the result of the human mind's tendency to see

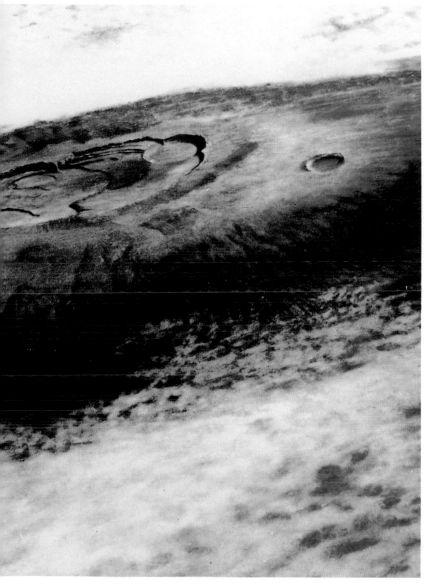

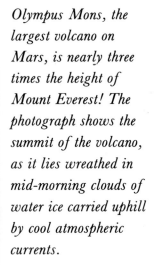

Olympus Mons, the largest volcano on Mars, is nearly three times the height of Mount Everest! The photograph shows the summit of the volcano, as it lies wreathed in mid-morning clouds of water ice carried uphill by cool atmospheric currents.

order in random features glimpsed dimly at the limits of the eye's resolution', as Abell, Morrison and Wolff aptly put it.

The largest planet in our solar system is Jupiter. It contains two-thirds of all the matter in the solar system that is not in the Sun. If it were eighty times more massive, it would become self-luminous and become a star in its own right!

Continues on page 22

*The Great Red Spot is some 25,000 km/16,000 miles long – large
enough to cover three Earths!*

*Jupiter has sixteen moons and is shown here with its four largest:
Io (upper left), Europa (centre), Ganymede and Callisto (bottom).
Ganymede and Callisto are comparable in size to the planet
Mercury.*

21

Jupiter's ring system appears to be associated with two tiny satellites or moons. These were first discovered by the Voyager spacecraft.

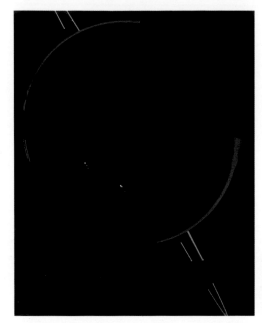

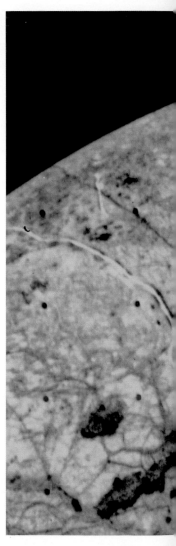

The mean distance of Jupiter from the Sun is 778 million km/486 million miles, and it has a diameter eleven times larger than that of the Earth. Its most prominent marking is the Great Red Spot. This spot had already been noted by the Italian astronomer Giovanni Domenico Cassini as early as 1665, but it became famous in the year 1878. It was described as 'very prominent and brick-red'. It is an area of much turbulence where hurricane winds with speeds of up to 360 km/225 miles per hour blow. The rim of the Great Red Spot circulates counter-clockwise once every six days.

Jupiter does not rotate like a solid, rigid body. It rotates 'differentially', which means that the period of rotation at its equator is not the same as its period of rotation at the poles. At equatorial latitudes, Jupiter completes a full rotation in only 9 hours and 50 minutes. Near the poles the rotation period is some five minutes longer. The first person to notice this was Cassini, who came to his conclusion by following features in the different Jovian belts.

Jupiter, like Saturn, has been found to be encircled by a ring of extremely fine particles. This ring was discovered by the Voyager space probes, and was never detected by earthbound telescopes. From the Voyager photographs the ring was found to be some 5,200 km/3,250 miles wide.

There is great diversity in terrain from one Jovian moon to the other. Europa is covered in water ice, while Io is the first body outside of the Earth found to have active volcanoes. Since eruptions continually resurface this satel-

The cracks in the water ice covering Europa are clearly visible in this Voyager photograph. The ice on this moon could be up to 100 km/63 miles thick. Below the ice the composition is presumably rocky and metallic.

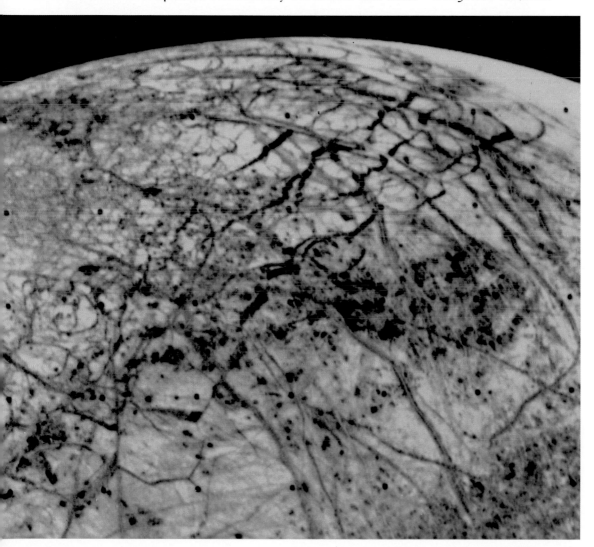

lite with sulphur and silicates no impact craters on this moon are to be found. Io has a diameter of 3,632 km/2,270 miles, similar to that of Europa and our moon, but it is geologically the most active body of our solar system.

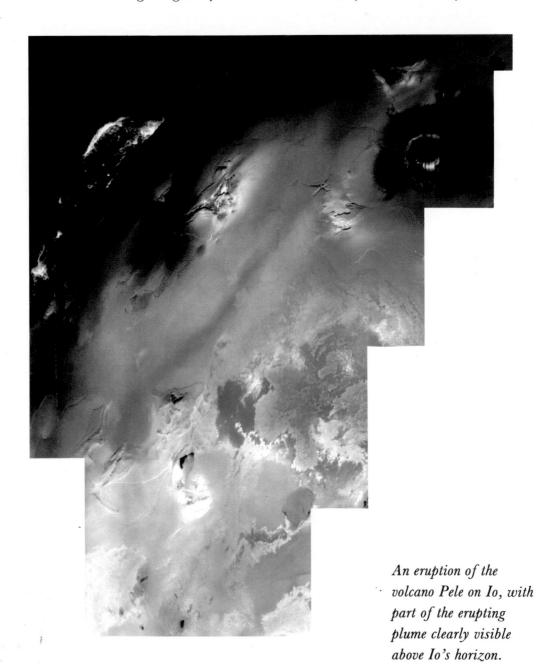

An eruption of the volcano Pele on Io, with part of the erupting plume clearly visible above Io's horizon.

Saturn, with its sets of rings, is spectacular!

The planet lies 1,427 million km/892 million miles from the Sun – nine and a half times the Earth-Sun distance. The planet has a diameter of 119,300 km/74,500

miles at the equator, and 107,700 km/67,300 miles at the poles. The time taken for Saturn to complete one turn at its equator is only 10 hours and 14 minutes, which is less than one half of our 24-hour Earth period. Considering that this

The rings of Saturn are much easier to see than Jupiter's ring. They were already observed in the 1600s by Galileo Galilei and Christiaan Huygens using their small and primitive telescopes.

planet has a diameter nine times that of the Earth, Saturn spins around on its axis very quickly.

If Saturn were to be placed in a gargantuan sea of water, the planet would float! This is because Saturn has a density less than that of water.

If a moon passes too close to a parent planet, gravity can cause the body to break up into millions of tiny pieces, forming a system of rings.

Surrounding the nucleus of a comet is a huge gas cloud, the coma, seen here in a photograph of Halley's Comet taken at the beginning of December, 1985. Also evident is the early formation of a tail.

The air was crisp and clean. And as I looked up toward the east, I saw this exquisite object blazing across the eastern horizon. Its head and tail were clearly visible. I set up my 35mm camera on a tripod, put in some high speed film and took what were to be my very first astronomical photographs. What a captivating experience to see such beauty, without even the aid of binoculars or a telescope! My fascination with astronomy began that morning of 22 March 1970.

Comets consist of a nucleus and coma (which together form the head of the comet), and one or more tails. In 1950, Harvard astronomer Fred L. Whipple proposed that the nucleus of a comet was a solid object, a few kilometres across, made of water ice mixed with silicate grains and

In February 1986 Halley's Comet reached its closest approach to the Sun. The comet has brightened considerably and the tail structure has changed dramatically. At least twelve separate tails can be distinguished and they extend for over three degrees (or six times the full moon's diameter) in the sky.

dust. His proposal became known as the 'dirty snowball' model.

Whipple's hypothesis was confirmed by recent observations of the famous Halley's Comet. Every seventy-six years this comet faithfully returns to our skies, the earliest reliable recording having been made by the Chinese in 240BC. During its 1985–86 return, the comet suddenly doubled its water evaporation rate. Continuous monitoring of the rate at which the ice was melting and evaporating as it approached the Sun showed a large jump from thirty tons per second on 19 February to sixty tons per second by the following day. Presumably a layer of dusty debris encasing some underlying ice had just blown off, giving credence to Whipple's 'dirty snowball' model for a comet's nucleus.

By March 1986 the comet was moving away from the Sun. The tail has grown to some ten degrees, or twenty times the full moon's diameter.

THE MISSING RECORD

The number of recorded returns for Halley's Comet from 240BC to the present was twenty-nine instead of the expected thirty. No mention of the reappearance in 164BC could be found. That is until a few years ago, when two Babylonian astronomical tablets housed in the British Museum were deciphered by experts Hunger and Walker. These tablets provided the missing evidence and established an unbroken record of sightings of this comet over the past 2,200 years.

'. . . the comet which previously had appeared in the East in the path of Anu in the area of Pleiades and Taurus, to the West . . . and passed along in the path of Ea.'

During March 1986 the European satellite Giotto penetrated to within 540 km/335 miles of the nucleus of Halley's Comet. A most important find was that the dust intermixed with the ice was very dark – 'darker than coal dust and more like black velvet', exclaimed Whipple. This meant that only a small percentage of the sunlight striking the nucleus was reflected. The nucleus was also discovered to be much larger than anticipated. Typical dimensions for the sizes of cometary nuclei were believed to lie in the range of one to five kilometres, but the Giotto spacecraft revealed a nucleus some sixteen kilometres long and about eight kilometres across. The nucleus was found to be very porous, causing astronomer Whipple to comment, 'Halley's nucleus must be one-third to nine-tenths empty space'.

Halley's Comet changes its appearance as it approaches, and then speeds away from, the Sun. As comets near the Sun, the lengths of their tail(s) increase. Cometary tails can be very long, extending as much as 150 million kilometres into space. But the amount of matter in the tail is exceedingly small. It is so rarefied that a comet has been described as 'the closest thing to nothing which can still be something'. Indeed, the tail is a much better vacuum than can be simulated in laboratories on Earth.

Large clouds of neutral hydrogen gas have been found to surround the heads of several comets. Comet Tago-Sato-Kosata was studied from outer space by the second Orbiting Astronomical Observatory in 1969, and was found to be surrounded by a hydrogen cloud approximately one and a half million kilometres in diameter. Similar clouds, invisible from the Earth yet visible from outside the Earth's

Continues on page 38

THE STAR OF BETHLEHEM

The appearance of a bright comet has been thought by some to herald important events. Shakespeare, in *Julius Caesar*, wrote,

'When beggars die, there are no comets seen;
The heavens themselves blaze forth the death of princes.'

The birth of Jesus Christ was heralded by a star in the sky, the exact nature of which is still not known. Some have speculated that a comet appeared in the heavens. Others calculate that a few years before the traditional date of Jesus' birth the planets Jupiter, Saturn and Mars lined up together in a unique way. Star gazers of the day would have noted this, but would they have described it as a 'star'? Yet others suggest that the Star of Bethlehem was a bright exploding star, but there are no records of a bright supernova at that time.

My personal belief is that it was an object supernaturally created by God for the event. No ordinary astronomical object would move in the way described in the nativity accounts, and if it were an ordinary and not a special star it surely would have been seen by King Herod and by everyone in Jerusalem and Bethlehem. Of one thing I am certain: on that great day, when the angels proclaimed *Gloria in excelsis Deo*, the heavens indeed declared the glory of God to the wise men in a unique way.

A panel in the Poor Man's Bible Window in Canterbury Cathedral portrays the wise men following the star.

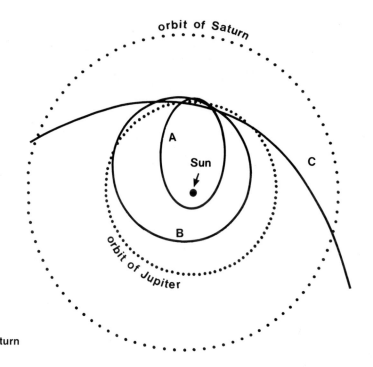

Lines A and B represent the old paths of Comet Lexell; the new orbit due to the gravitational effect of Jupiter is shown as C. The dotted ellipses indicate the orbits of Jupiter and Saturn.

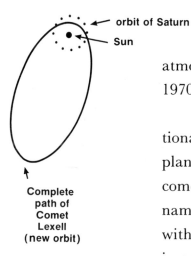

atmosphere, were found to surround Bennett's comet of 1970 and Kohoutek's comet of 1973.

The more massive a planet, the stronger its gravitational force. Jupiter, and to a lesser extent the other giant planets, can radically alter the orbits of some comets if the comets happen to pass very close to them. One such comet, named Comet Lexell, suffered two very close encounters with Jupiter in the years 1767 and 1779. A dramatic change in orbit occurred! Whereas Comet Lexell would once have taken just under six years to complete an orbit about the Sun, the massive planet Jupiter has so affected its path that it now requires over 260 years to complete one revolution!

VIEWPOINT EARTH

SOME OF THE LARGEST AND MOST
SOPHISTICATED OPTICAL TELESCOPES
IN THE WORLD HAVE BEEN USED TO
SECURE THE COLOUR AND BLACK-AND-
WHITE PHOTOGRAPHS IN THIS BOOK.

As you turn these pages you are looking through telescopes
such as the famous Hale telescope in California, the giant
Anglo-Australian telescope in Australia, and the towering

*Modern telescopes
dwarf the astronomers
who use them. Their
light-gathering power is
measured by the
diameter of the light-
reflecting mirror. This
photograph shows the
200-inch Hale telescope
on Mount Palomar,
California.*

The Anglo-Australian telescope located at Siding Spring, Australia.

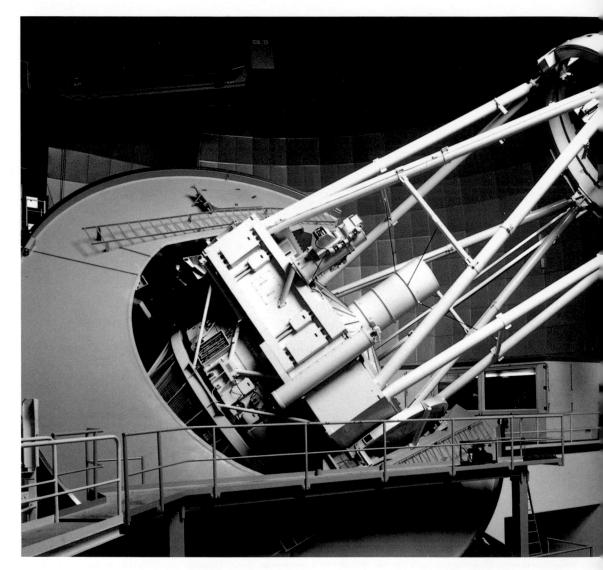

Mayall telescope at Kitt Peak, Arizona.

When making photographic exposures, astronomers invariably operate these telescopes at the 'prime focus' configuration. This means that the observer sits in a cage suspended at the very top of the telescope tube collecting as much reflected light as possible from the mirror below. Dr

The famous astronomer Edwin Hubble is seen here using the Mount Palomar telescope at the prime focus position. ▷

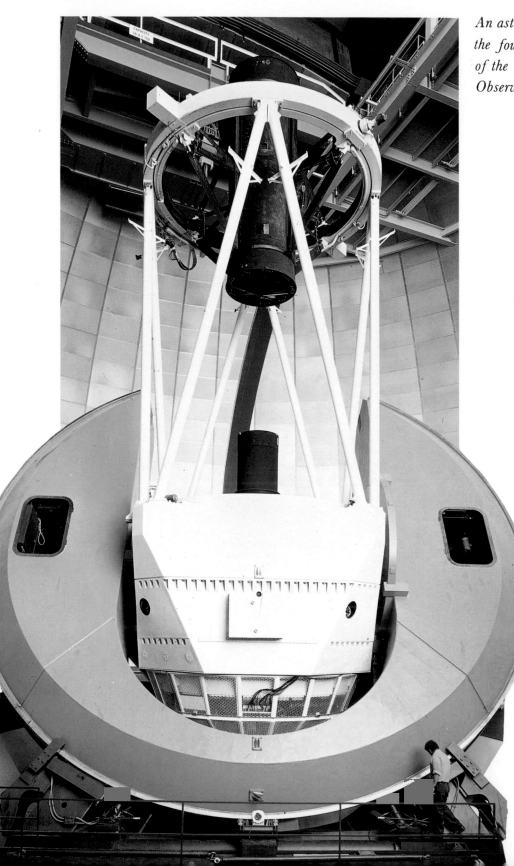

An astronomer operates the four-metre telescope of the Peak National Observatory.

The 1.2 metre United Kingdom Schmidt Telescope. This equipment was used to take several of the photographs in this book.

David Malin, one of the foremost experts in astrophotography in the world today, used the prime-focus configuration at the Anglo-Australian Telescope to obtain many of the photographs we shall be discussing. Dr Malin sat inside the telescope working at a height of several storeys above ground level!

While such giant telescopes can be used to detect extremely faint objects almost at the very boundary of our observable universe, their fields of view in the sky are very limited. Only a tiny portion of the sky can be viewed at any one time. For this reason several of the photographs were taken with wide-field Schmidt telescopes, such as the highly sophisticated United Kingdom Schmidt Telescope in Australia, and its twin in the northern hemisphere, the 1.2-metre Schmidt at Mount Palomar.

Other views are just not available from Earth's surface. Only spacecraft can supply these – a Martian sunrise, for example, or the volcano Pele erupting on one of Jupiter's moons. We are fortunate to be living in an era when breathtaking close-ups of some of the planets and their moons are no longer in the realm of science fiction. Highly successful space probes such as Voyager and Viking have sped past or landed on these planets, and have transmitted back to Earth startling photographs. A few of these views are included in this book.

The Pleiades cluster is a magnificent open cluster of stars in the constellation of Taurus. To the unaided eye, only six or seven stars may be seen, and for this reason the name 'Seven Sisters' is sometimes used for this group. But the cluster is actually a collection of several hundred stars. The twenty brightest of these stars are blue-white and very hot. Extensive clouds of dust and gas surround and reflect the light from the embedded stars, producing the cloudy effect around the brighter cluster members.

Photographs through large telescopes can be misleading, in that they can make extended far-off objects appear to be relatively nearby. Star clusters, such as the Pleiades with its wispy nebulosity, might appear to be our near

The light now reaching us from the Pleiades cluster set out 400 years ago.

neighbours in space. And yet the distance even to the closest star outside the solar system, Proxima Centauri, is staggeringly large. So much so that expressing the distance in kilometres or miles would make 'millions' and 'billions' the most common word in this book! The Pleiades cluster of stars, for example, is 4,000 million million kilometres away! Large numbers like these are best expressed in terms of 'light years'. Light or radio travels incredibly fast, and a radio message can cover the distance between us and the moon in little more than a second. In a year light will travel roughly 10 million million kilometres and so the Pleiades cluster may be said to be 400 light years away.

COSMIC CRADLES

A view of the entire Rosette Nebula (overleaf) and a close-up of part of the same nebula (left) highlighting the Bok globules. Because these globules produce no light of their own, they are only seen as tiny dark dots silhouetted against the nebula itself. The diameters of the smaller globules are only 100 times the diameter of our solar system.

IN SOME OF THE PHOTOGRAPHS IN THIS SECTION, WE ARE LOOKING AT STARS IN THE PROCESS OF BEING BORN.

How are stars formed? What is the source of their brilliant light?

Young stars are invariably found in the clouds of gas and dust which lie in interstellar space. These glowing gaseous clouds are called nebulae.

The likely birth places of stars within these nebulae are little dark globules of gas and dust, called Bok globules

Continues on page 50

after the Dutch-American astronomer B.J. Bok. Protostars may be formed by the contraction, under gravity, of these Bok globules, until eventually the temperature becomes high enough for nuclear fusion to start, and a proper star is formed.

A striking example of this is seen in the Cone Nebula, which contains a very young cluster of stars. Among the cluster members are young, low mass stars called T Tauri stars which are actually believed to be in the process of contraction under gravity to form 'proper' stars, where hydrogen can start burning in their cores. The Cone Nebula lies in the constellation of Monoceros and is 2,600 light years away from the Earth. Looking at the photograph of the Cone Nebula, I am reminded of some words once written by author Timothy Ferris: 'Stand under the stars and say what you like to them. Praise or blame them, question them. The universe will not answer. But it will have spoken.'

Nebulae provide astronomers with a means of studying the basic properties of gases under conditions simply unobtainable on Earth. The Great Orion Nebula, although so spectacular, is nearly a vacuum! It contains only 600 atoms per cubic centimetre, whereas the density of air at sea level on Earth is:

10,000,000,000,000,000,000 atoms per cubic centimetre!

Not only can the Orion Nebula, otherwise known as Messier 42, be studied at optical wavelengths but it is also a source of both X-rays and infra-red rays. The infra-red sources are evidently associated with the protostars – very young stars in formation, still wrapped deep in their cocoons of dust. In Messier 42, the site of on-going star

Continues on page 54

A young star cluster born here is still connected with the dense cloud of interstellar matter from which it was formed – notice the long 'finger' of dark matter pointing upward from below, hence the name 'Cone Nebula'.

From Radio Waves to X-rays: The Electromagnetic Spectrum

Drop a stone into a pond and the ripples will spread outward in ever increasing circles. The size of the ripples, and the distance between one ripple and the next, are determined by the size of the stone: a boulder will form bigger waves than a pebble!

Electrical radiation in space can also be regarded as a series of waves. An observer with suitable equipment would be able to measure passing waves of electrical and magnetic energy, rising and falling with a particular rhythm. But whereas the ripples in the pond

Electricity supply   Audio   Radio waves

may lap the shore only two or three times a second, the frequency of electromagnetic waves may be many millions of 'laps' per second.

The human eye is sensitive to particular frequencies of this radiation. If the waves reach us at a rate of around 600,000,000,000,000 cycles (or 'laps') a second we can see them! This is light, and the different colours represent different frequencies. Waves of higher frequency we call ultraviolet radiation and X-rays, and while we cannot see them we can build equipment to detect them. Waves just below the visible spectrum we call infra-red radiation, and still lower frequencies we use for television and radio waves. Different stars emit electromagnetic radiation at a variety of frequencies, and radio telescopes are used to detect these waves if they are below the visible spectrum.

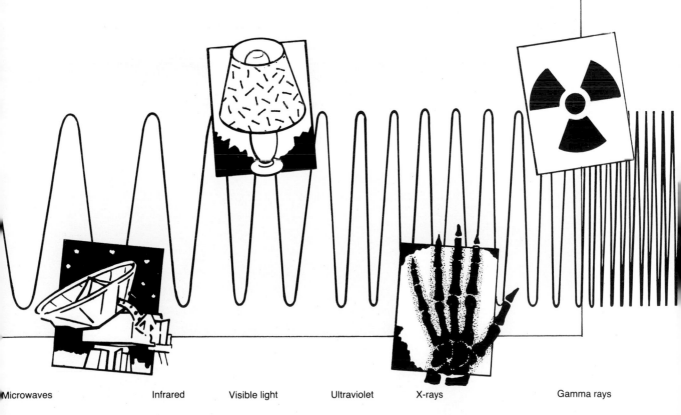

| Microwaves | Infrared | Visible light | Ultraviolet | X-rays | Gamma rays |

formation actually lies in giant clouds of molecules located behind the visible nebula. Under certain conditions, atoms in an interstellar gas cloud can link up to form molecules and the molecular clouds can then break up, in due course, to form stars.

Continues on page 58

A more complete view of the Cone Nebula with its surroundings. The Cone Nebula itself lies to the extreme right of the photograph. The presence of much gas in the entire area is obvious. The very bright, conspicuous star at the extreme left edge of the photograph is called S Monoceros.

The majestic Orion Nebula is thirty light years in diameter and at the centre of this nebula lie four young stars, known as the Trapezium. These stars cause the nebula to glow. ▷

The Horsehead Nebula. The 'horse's head' lies to the right of the centre area of the photograph. The very bright star is Zeta Orionis. The cross effect is not real but only an artefact of the optical system used to take the photograph.

A close-up of the Horsehead, taken through one of the world's largest telescopes. The dark, prominent head measures one-third of a light year by one light year, and the neck below the head is one light year across.

The Horsehead Nebula is part of a magnificent and complex gaseous region in the constellation of Orion. It lies at a distance of some 1,600 light years from us and gets its name from the striking feature of a very prominent dark nebula shaped in the form of a horse's head. Embedded in the large dust cloud below the Horsehead are two bluish patches, called reflection nebulae. These nebulae shine by reflecting the light of stars located within the dusty clouds.

The Horsehead is not simply a quiescent region of gas and dust, as one might at first suppose. It is an active site, where stars of low mass are continuing to be formed. This nebula is a very fine example of a dark globule, which, as we have seen, is often a site of star formation. There is so much gas and dust present that if the brightest star in our night sky, Sirius, were to be placed immediately behind the Horsehead, not only would Sirius fade from visual detection with the naked eye, but we would need the largest of telescopes to photograph our otherwise brightest night-time star!

The Lagoon Nebula is a stellar nursery. It lies 4,000 light years from us, and measures thirty light years across.

The Lagoon Nebula, Messier 8, in Sagittarius, is just visible to the naked eye. Not only are large amounts of cosmic dust responsible for producing the distinct 'lagoon' shape, but they are also the source of the large numbers of Bok globules scattered throughout the nebula. The Trifid Nebula derives its name from the fact that it appears to be divided into three major sections by prominent lanes of interstellar dust. The nebula contains several dense

Continues on page 62

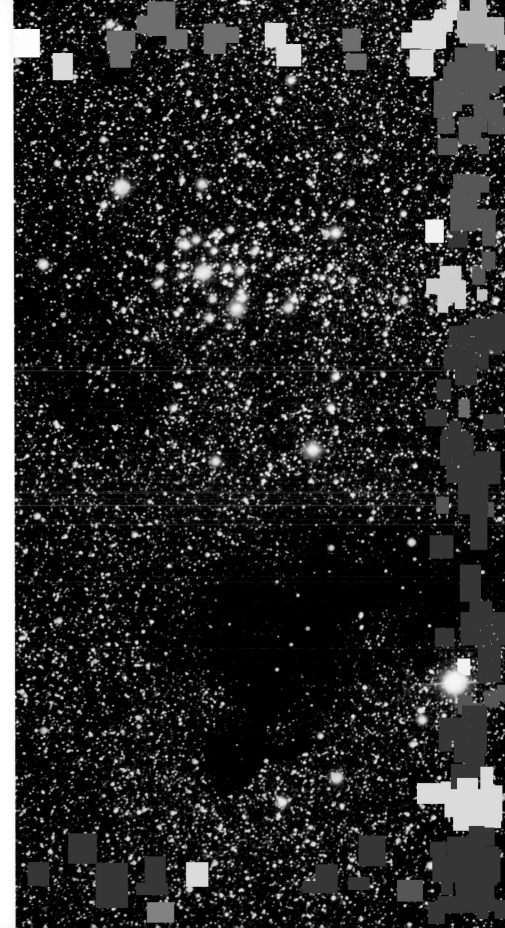

◁ *The Trifid Nebula,*
another stellar nursery,
is 3,500 light years
away.

A dark cloud of
interstellar dust, this
time bearing no
resemblance to any
familiar shape, may be
seen in Sagittarius.
Myriads of stars are
visible in the
photograph, and the
only reason that we can
see the dark globule
(called Barnard 86) is
because it blocks out the
light from the stars
beyond it.

globules of dust within the softish, orange areas. Some of these might be in the process of contracting to form new stars.

Who am I, in comparison to such majesty seen in the heavens? In purely physical terms I am dwarfed into insignificance.

But I am created in God's own image, and so I am not a nobody, but a somebody. I am unique and irreplaceable.

This very bright star is called S Monoceros.

Colours of the Stars

There is a very wide range in the colours of stars. There are cool stars that radiate with a deep red colour, and hot stars that radiate with a deep blue colour. A star's colour indicates its surface temperature – in other words, how hot it is.

When an iron rod is placed in a fire, it first glows red, then white, as the temperature in the rod increases. And so it is with stars. The blue stars are much hotter than those of a yellowish colour. Whereas the surface temperature of our Sun is 6,000 degrees Centigrade, the hottest of stars can have surface temperatures exceeding 35,000 degrees Centigrade! The surface temperature of cool red stars may be less than 3,500 degrees Centigrade, although lower surface temperatures are possible in protostars which are stars in the making.

Astronomers group stars in a sequence of decreasing temperature from O to M:

O   B   A   F   G   K   M

Rigel, for example, in the constellation of Orion, is a very hot blue star, of type B. Betelgeuse, also in Orion, is red, and of type M. Yellow stars, such as our Sun, belong to class G.

The range in the diameters of stars is enormous. There are white dwarfs smaller than the Earth, to supergiant stars whose diameters can exceed several hundred times that of the Sun. If all the stars could be viewed at the same distance without any obscuring

What makes some stars shine blue and others red? The colour of the light shows how hot the star is at its surface.

interstellar dust, the star Eta Carinae would outshine the Sun by more than six million times. The least massive known star, called RG 0058.8–2807, would have a total luminosity only a ten-thousandth that of the Sun!

OUR OWN STAR

THE NEAREST STAR, AND THE ONE WITH WHICH WE ARE BEST ACQUAINTED, IS THE SUN.

'Sun, moon and stars are God's travelling preachers.' C.H. Spurgeon.

Our closest star, the Sun, is a blazing ball of hydrogen gas. At the very core of the Sun, 584 million tons of hydrogen are being converted through nuclear processes into 580 million tons of helium each second. This means that the Sun is continually 'losing' mass at a rate of four million tons *per second*. This mass loss in the core is released in the form of energy as light and heat, without which we on Earth

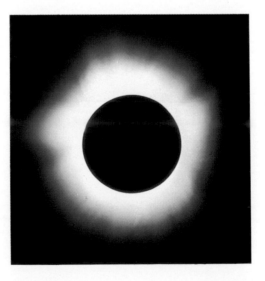

A solar eclipse. The temperature at the centre of the Sun is exceedingly high: some 16 million degrees Centigrade. At the surface it is much lower: about 6,000 degrees Centigrade.

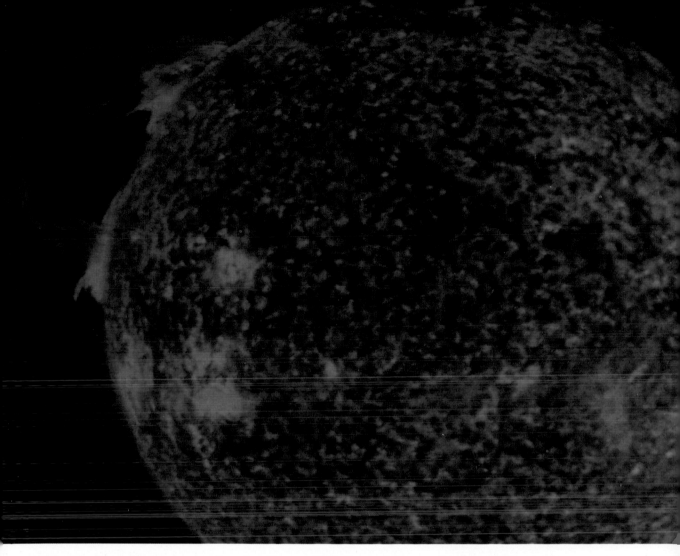

could not live. But eventually the Sun will fail to shine. Several years ago I mentioned this to an audience, and a very worried voice piped up from the back,

'Doctor, how long have we still to go?'

'There is provision for another 5,000 million years,' I replied, and his face beamed.

How remarkably well placed is the Earth's orbit around the Sun! If it were too close to the Sun, we would burn . . . or too far away, freeze. This is not by chance or accident. God in his infinite wisdom and power made it so: 'God made the two larger lights, the sun to rule over the day and the moon to rule over the night; he also made the

A solar eruption can reach altitudes above the solar surface of up to 500,000 kilometres, which is forty times the diameter of the Earth!

stars.' (Genesis, chapter 1).

What a thrilling experience it is to use special equipment to photograph the seething surface of the Sun! Solar prominences . . . sunspots . . . flares . . .

Prominences are large gaseous eruptions on the surface of the Sun. Quiescent prominences can hover over the Sun for weeks or months, while eruptive prominences change their appearance over a matter of hours.

If a solar prominence is seen by an observer on Earth at 4.03 p.m., he or she must remember that the eruption is not actually taking place at that moment but that it happened just before 3.55 p.m. In looking at the Sun, we are looking back a little over eight minutes in time. This is true in all of astronomy. We are continually looking back minutes, hours, years, hundreds of years . . . millions of years . . . billions of years into the ever distant past. Light does not travel infinitely fast, but at a limited speed of 300,000 km/186,000 miles per second. In the case of our Sun, light leaving the solar surface reaches our eyes just over eight minutes later.

Also to be seen on the surface of the Sun are 'sunspots'. The central inner region of a sunspot is called the umbra, and is surrounded by an outer and lighter penumbra. These areas look darker in colour than their surroundings simply because they are cooler in temperature – around 4,000 degrees Centigrade instead of 6,000 degrees Centigrade. Every eleven years the number of sunspots reaches a maximum, followed at the half-cycle by a sunspot minimum. Radio reception on Earth is affected by sunspot activity, particularly if the Sun is approaching a sunspot maximum.

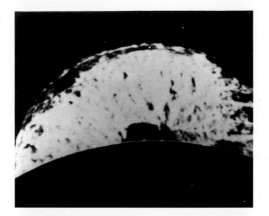

One of the most
spectacular eruptive
prominences ever seen
occurred on 4 June
1946. At 4.03 p.m. the
prominence took the
form of a huge arch.

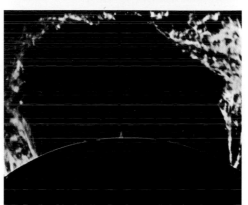

4.36 p.m., 33 minutes
later: the prominence
was at maximum
development.

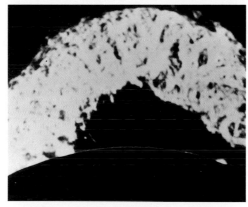

By 5.03 p.m. the arch
had become much more
tenuous, but still
extended up to 320,000
kilometres, twenty-five
times the Earth's
diameter.

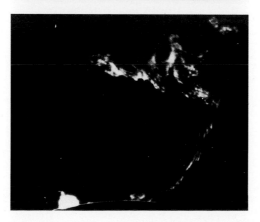

5.23 p.m., 1 hour and
20 minutes later: little
remained of the great
arch as most of the
hydrogen gas had
dispersed into space.

Our closest star is thus very active. At the surface violent short-lived solar flares can release the equivalent of two billion megatons of TNT. And at the centre 584 million

SUNSPOTS AND RADIO RECEPTION

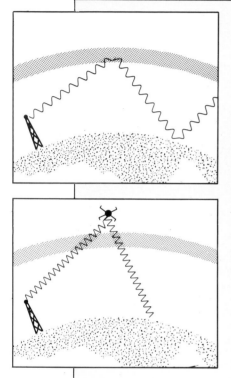

The high-energy radiation from the Sun maintains bands of charged atoms and molecules around the Earth. These layers, which form the 'ionosphere', extend outwards from about 75 km above the Earth's surface and are important for radio communications. The ionosphere reflects radio waves back down to the surface. Certain wavebands are more easily reflected than others and signals on shortwave radio bounce between the ionosphere and the Earth so enabling them to travel around the globe. It is like using a mirror to see around corners!

Television signals usually pass directly through the ionosphere and into space, and so long-distance TV communication requires a satellite to receive the signals and re-transmit them back to Earth.

The ionosphere is always moving and it changes considerably during darkness. Radio conditions are liable to fluctuate, and signals from a distant transmitter may fade away at certain times of day. When the Sun is at its 'sunspot maximum' shortwave radio conditions are at their best and radio 'hams', using only transmitters as powerful as a domestic light bulb, are able to communicate with fellow enthusiasts around the world.

tons of hydrogen fuses into 580 million tons of helium second by second.

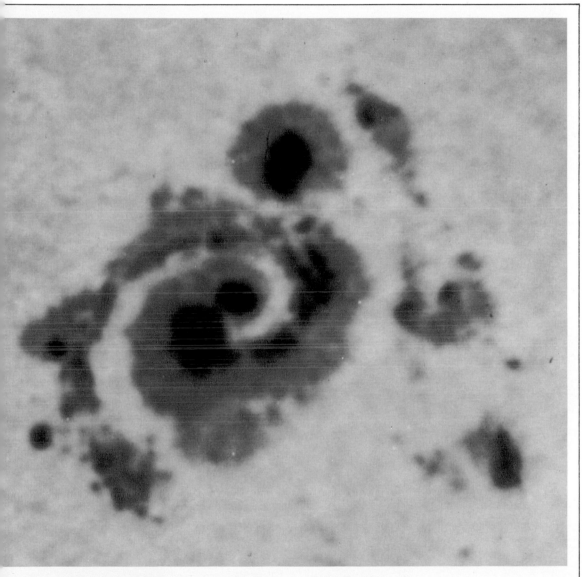

An average sunspot is about twice the diameter of the Earth, and may be seen on the Sun's surface for about one week. Sunspots generally occur in groups and these groups last for longer periods, possibly up to two months.

THE POWER OF THE SUN

Hold a stone above a pond of water and let it fall. The Earth attracts the stone and it accelerates downwards, hitting the water with a splash. This everyday experience demonstrates what we call gravity, and that the *gravitational force field* is attractive and not repulsive – stones do not fly upwards! In the splash one form of energy is converted into another, for some of the energy of the moving stone is changed into sound.

Small atomic particles called protons (which are one of the building blocks of matter) repel each other because of their electrical charge. But if the pressures in a gas are extremely high, as in a core of a star, they may move toward one another and a force known as the *strong nuclear force* will overcome the electrical force driving them apart. The result is that protons, and neutrally charged particles called neutrons, are held together much as gravity holds a stone at the bottom of a pond. When protons and neutrons fuse together under this nuclear force, vast amounts of energy are released. The generation of energy in the cores of stars such as our Sun is by means of such nuclear fusion. Four atoms of the simplest element hydrogen combine to form one atom of helium.

Since hydrogen is the most common element in the universe why does the fusion reaction not occur all around us – in clouds or in the sea? Fortunately for us the strong nuclear force operates over a very short distance, and positively charged protons are normally

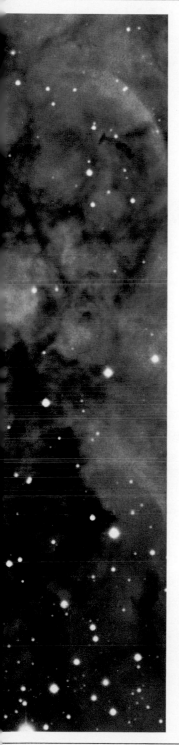

repelled and move away from each other long before they can come close enough to be bound together by the nuclear force. What is needed to drive the fusion process are very great pressures and temperatures. The core of a star provides just the right conditions for the process to begin. Then, as fusion energy is released, sufficient excess energy is provided to force more particles to combine and the whole process becomes self-sustaining.

Our own star, the Sun, only appears bright because we are very close to it. But if all the stars were viewed at the same distance, the star Eta Carinae (left of centre) would outshine the Sun by more than 6 million times.

Supernovae and the White Dwarfs

In the upper photograph, the white dwarf is lost in the glare of Sirius; it can only be seen in the lower exposure. The white dwarf Sirius B has a mass very nearly the same as the Sun's, but in size it is smaller than Earth!

SOMETIMES STARS EXPLODE. SOMETIMES THEY CONTRACT. SOMETIMES THEY SPIN AT GREAT SPEED. WHAT MAKES THESE THINGS HAPPEN?

Nuclear fusion in the cores of stars cannot continue for ever! As there is only a limited amount of the supply gas hydrogen available, these immense fireballs will eventually burn up. In the Sun's core, for example, 584 million tons of hydrogen are converted into 580 million tons of helium every second. And this can only continue for a fixed period of time.

After a star has exhausted its nuclear fuel, it can at first expand, but will eventually collapse, as there is no longer any source of energy to hold the star up against the force of gravity dragging everything inwards. Atoms in the collapsing gas cloud become squeezed closer and closer together, and electrons become dislodged from their orbits about their nuclei. At this stage there is a floating sea of electrons. And since each holds a negative charge of electricity the electrons repel each other, just as like poles of magnets repulse each other.

If the mass of the star in the process of gravitational

74

collapse is comparable to the mass of our Sun, the pressure from these electrons can become sufficient to halt the contraction. Such a star is called a 'white dwarf' and the first white dwarf to be discovered was found alongside Sirius. Sirius, the brightest star in the night sky, is actually a double star and it is the faint companion, Sirius B, almost lost in the glare of Sirius itself, which is a white dwarf. White dwarfs are usually smaller than the Earth and support themselves against gravity by the repulsion pressure of 'degenerate electrons'.

But if the mass of the collapsing star moderately exceeds that of our Sun, the repulsion pressure from the electrons is not strong enough to stop continuing contraction. As the pressure in the star increases the electrons collide violently with the nuclei. During such collisions, the negatively charged electrons combine with the positively charged protons in the nucleus to produce 'neutrons'. Provided the mass of the original star is not too large (less than about three times the Sun's mass) contraction can still be halted, but the star will explode, producing a supernova with a superdense neutron star at its centre.

One teaspoonful of matter from a neutron star would weigh one billion tons! And a typical diameter of such a star is only ten to fifteen kilometres.

The Crab Nebula is the remnant of an exploding star, or supernova, which could be seen during broad daylight from Earth during the year AD1054. Since the Crab Nebula is 6,000 light years away, the actual explosion did not occur in AD1054 but 6,000 years earlier. Observations show that the nebula is expanding at speeds of up to 1,450 kilometres per second, and changes in the structure of its filaments can

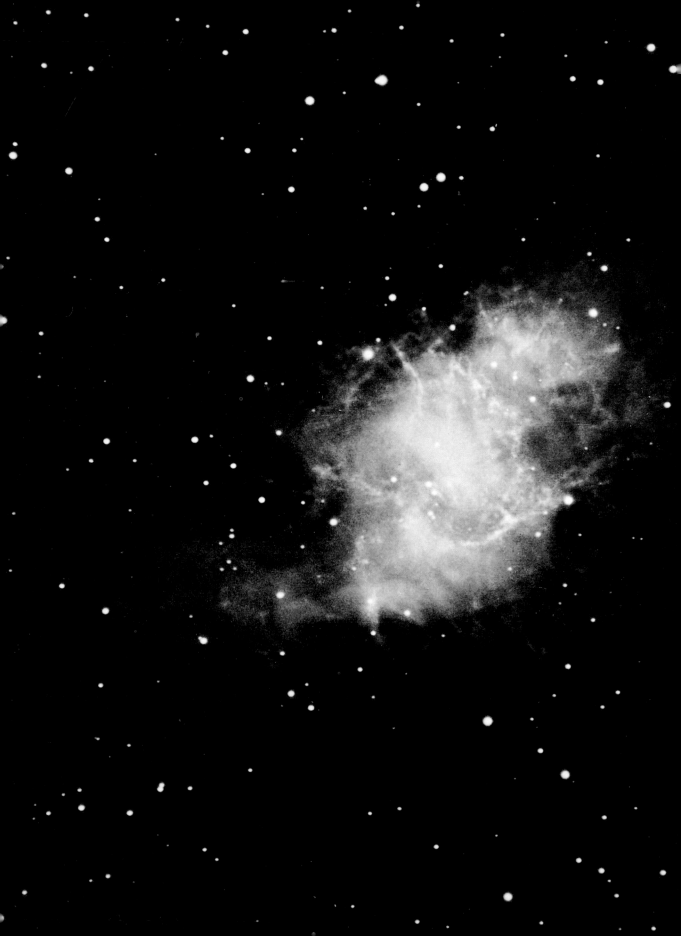

The Crab Nebula, so named by the astronomer Lord Rosse in 1848, is perhaps one of the most interesting and most studied objects in the sky.

be seen over time-scales of weeks.

At the heart of the Crab Nebula lies a superdense neutron star, spinning around on its axis thirty times every second. This neutron star has one of the shortest rotation periods known, although the fastest neutron stars have been found to spin around at an amazing rate of over 600 times per second!

One of our most basic needs is security. World news causes many to panic and to fear that either they, or the world, are about to go out of control. But the God who regulates the birth and death of stars, and the entire universe, is well able to give us all the security we need! Of this God the ancient psalmist exclaimed,

'He determines the number of the stars
and calls them each by name.
Great is our Lord and mighty in power;
his understanding has no limit.'

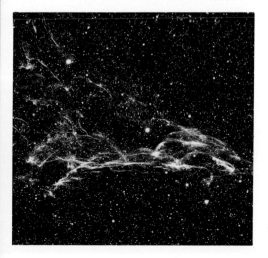

The central section of the Veil Nebula as seen through the 4-metre telescope at Kitt Peak.

telescope they bear a superficial resemblance to planets. In fact planetary nebulae have nothing whatsoever to do with planets. In size they are thousands of times larger than the entire solar system. A typical planetary nebula might have a diameter of a half to one light year across.

How do these fascinating planetary nebulae form?

A low mass star reaching the end of its nuclear fusion reaction in its core may first expand for a time before collapsing inwards. Hydrogen in the central core of the star becomes exhausted, but hydrogen gas can continue to burn in a thin interior shell around the core. This causes the star to expand – so much so that it becomes a bloated giant where the outer layers of gas simply escape into space. The inner core continues to contract leaving a planetary nebula, such as the great planetary nebula NGC 7293 in Aquarius.

The central star of a planetary nebula is exceedingly hot, but it is also very compact. Estimated temperatures range from 20,000 to over 100,000 degrees Centigrade, while many of them have diameters no larger than those of white dwarfs! These intensely hot and compact stars radiate intense ultraviolet energy which makes the gas shells around the star fluoresce. In other words, the shells absorb the ultraviolet rays from their central stars and re-emit this energy as visible light, so making these shells visible to our eyes.

Planetary nebulae never last long: an age of about 20,000 years is typical. A planetary nebula may be the last

The Great Planetary Nebula in Aquarius is the gas thrown off by an expanding star before it finally collapsed in on itself.

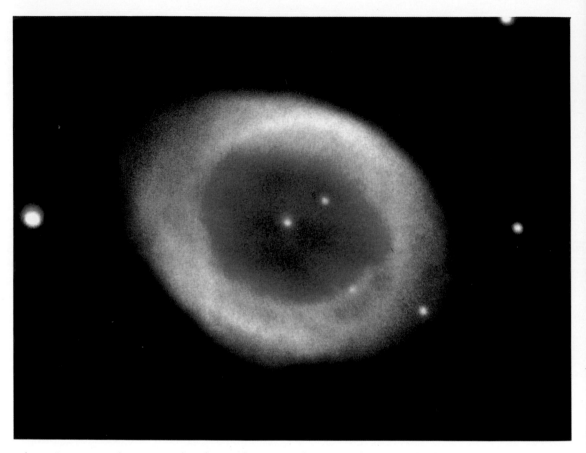

Amazing to ponder . . . that the compact central object was once the actual core of a star!

ejection of matter by certain stars before they collapse to form white dwarfs.

Ten times further away than NGC 7293 lies Messier 57 – a famous planetary nebula in the constellation of Lyra which is appropriately called the Ring Nebula. How did this nebula form? A low-mass star, such as our Sun, expanded and expanded, reaching the red supergiant phase. The star then shed its outer gaseous envelope, exposing its inner core. Now on its way to becoming a white dwarf, the central star still supplies the necessary ultraviolet radiation for the entire Ring (which is over one light year across) to fluoresce.

HOLES IN THE UNIVERSE

BLACK HOLES WITH SUCH INTENSE GRAVITY THAT NOTHING CAN ESCAPE – WE CANNOT SEE THEM SO HOW DO WE KNOW THEY EXIST?

White dwarfs and neutron stars are two possible end products of a collapsing star. The third possibility is where continued contraction can never be halted and the star forms a black hole.

One of the most exciting predictions of Einstein's general theory of relativity is the existence of black holes. Here gravitational forces become so intense that they prevent the escape of anything – even particles moving with the speed of light! This could happen when a very massive star reaches the end of its life cycle and collapses in on itself to such a degree that gravity overwhelms all other forces. The Earth would have to be squeezed down to a radius of less than one centimetre if it were to become a black hole. And the Sun would need to be compressed to a radius of about 3 km/1.8 miles!

A black hole is a region of space into which a star (or collection of stars or other bodies) has collapsed and from which no light, matter or signal of any kind can emerge. The terminology 'hole' is most appropriate, for the star

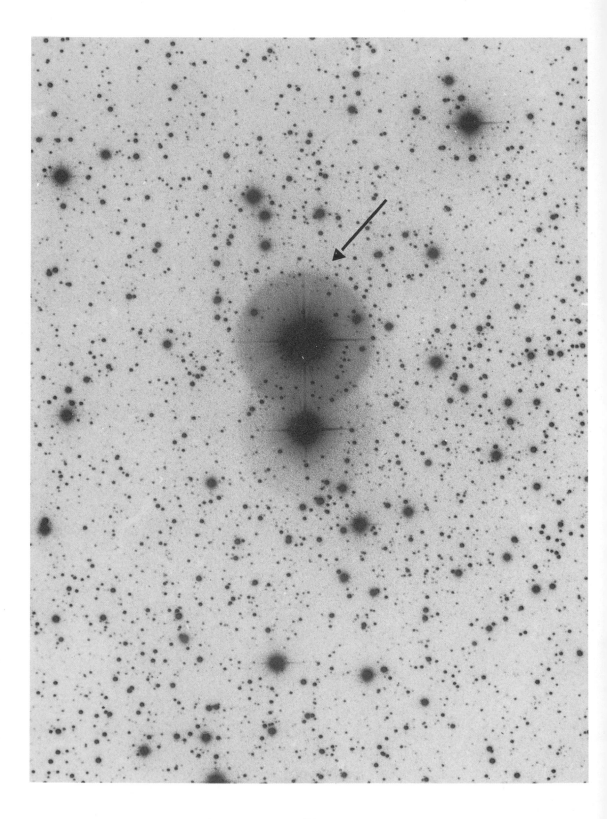

collapses ever inwards leaving a volume of space where gravity is strong enough to prevent anything from escaping.

Physicists distinguish two major types of black holes according to whether or not they rotate. The surface of a black hole is called its event horizon. Only events occurring outside this surface can reach an outside observer, for events inside the horizon can never influence the exterior. The event horizon of a non-rotating black hole is spherical in shape, while rotating black holes have horizons which are flattened at the poles, just as rotation slightly flattens our Earth at its poles. When a black hole first forms, its event horizon may have a grotesque shape and may be rapidly vibrating. But within a fraction of a second the horizon settles down to a unique smooth shape.

To the physicist and mathematician, black holes are remarkably simple objects. They are said to have 'no hair', meaning that any black hole can be uniquely characterized by very few parameters, or numbers. A non-rotating black hole, for example, is uniquely characterized by only one number – its mass. At the centre of a black hole lies an area where all known laws of physics break down. This is a region where matter can literally be crushed out of existence.

While black holes are predicted by Einstein's theory, have they ever been detected? The answer is a tentative yes. One promising candidate is found in the constellation of Cygnus and is called Cygnus X-1. Cygnus X-1 is an X-ray source, which optically coincides with the position of a blue

The arrow points at HDE 226868 – is this star circling about an invisible black hole?

85

'supergiant' star called HDE 226868. To us the star does not appear to be bright as it is some 6,500 light years away. Observations through a spectroscope show that this blue star is orbiting around an *invisible* companion, which is very compact yet has a huge mass. The X-rays could be produced from gas circling around the compact object. Gas from the blue star is sucked off and then circles around, before disappearing into what is in all likelihood a black hole.

Not all black holes result from the collapse of very massive stars. In the early history of our universe, certain regions may have become so compressed that they too underwent gravitational collapse to form black holes. Such small black holes are called primordial black holes, and a primordial black hole with the mass of a mountain would have the size of an atomic particle!

Artist's impression of Cygnus X-1 binary black hole system. It shows matter from a neighbouring star being sucked into the hole.

ALBERT EINSTEIN 1879–1955

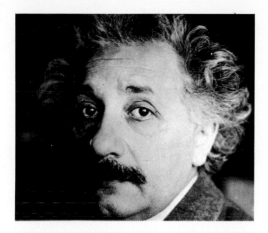

Albert Einstein was a German-born physicist who received the Nobel Prize in 1921 for his work. In 1905 he published his 'special theory of relativity' in which he revealed his famous equation: $E=mc^2$. This formula expresses how energy (E) is related to mass (m) by twice multiplying by the speed of light (c). The conversion of mass into energy is now understood as the source of the light and heat radiation of the stars, and is the basis behind nuclear reactors here on Earth.

The General Theory of Relativity, published ten years later, shows how gravity makes space become 'curved' so that light no longer travels in straight lines. If the gravitational force is strong enough it will pull the light back on itself. Einstein's scientific papers suggested a number of experiments by which his theories could be tested and the verification of his predictions has established relativity as a fundamental concept within physics.

THE MILKY WAY

As a familiar sight to many on dark moonless nights, the plane of the Milky Way stretches as a majestic celestial arch from horizon to horizon.

THE MILKY WAY CONTAINS 100,000 MILLION STARS – SO LARGE A NUMBER THAT COUNTING ONE STAR EVERY SECOND WOULD TAKE 2,500 YEARS!

To many who study the universe, galaxies offer a life-lasting intrigue and challenge. Their vastness, their complexity and their initial formation all pose many questions for astronomers today.

A galaxy is a large aggregate of stars and interstellar matter, bound together by gravity. The galaxy to which our Sun belongs is called the Milky Way, and is 100,000 light years in diameter. It contains all the objects we have so far discussed: the Pleiades star cluster, the Crab Nebula, the Veil Nebula and the Cone Nebula.

We cannot of course see our Milky Way from the 'outside', for we are part of it. But we can look beyond the stars of our own galaxy to other 'Milky Ways'. The shapes of these galaxies fall into three major classes: elliptical, spiral, and irregular. On photographs, an elliptical galaxy looks like a football or, more often, like a squashed football bulging at the centre. Spiral galaxies are lens-shaped and

Some of the estimated 100,000 million stars of the Milky Way, with Halley's Comet in the foreground. ▷

Andromeda resembles our own Milky Way, although its diameter is slightly larger. Light from the spiral, travelling at 300,000 kilometres per second, has journeyed for over 2 million years before reaching the camera lens! As observers on tiny planet Earth we can feel very insignificant given the immensity of it all.

It took the equivalent of about six weeks of observing time with South Africa's largest optical telescope to secure this remarkable photograph of the very centre (shown arrowed) of our Milky Way. With the help of special infra-red equipment constructed by astronomer Ian Glass, we are looking through some 26,000 light years of interstellar gas and dust. After computer processing the image, we can see with amazing clarity an otherwise completely inaccessible region of our galaxy.

possess luminous spiral arms which consist of young stars, dust and gas. The Andromeda Spiral, or Messier 31, provides a nice example of a spiral-shaped galaxy. Irregular galaxies do not present any regular spiral structure or central bulge.

Galaxies generally lie in small groups or in larger clusters which contain thousands of member galaxies. Our Milky Way does not lie in a cluster, but in a group of about thirty members. This is known as the 'Local Group' and is some 10 million light years across. The Local Group contains three spiral galaxies, the Milky Way, the Andromeda Spiral, and Messier 33. The Andromeda Galaxy is 2.2 million light years distant, and is the nearest spiral galaxy to our own, while Messier 33 lies at a distance of some 3 million light years.

The Large Magellanic Cloud is our closest galactic neighbour and lies at a distance of 150,000 light years, fifteen times closer than the Andromeda spiral.

The Local Group contains four galaxies of the irregular type, the largest of which is the Large Magellanic Cloud. In the cloud is the famous Tarantula Nebula, which is a giant region of ionized hydrogen gas 780 light years across. During the early months of 1987 every large optical

Continues on page 96

The brightest supernova in our skies since 1604! The photograph shows the Tarantula Nebula with the star in its 'pre' and 'post' explosion phases. An arrow pin-points the supernova. Of course this star actually exploded 150,000 years ago, but the light has taken thousands of years to reach us.

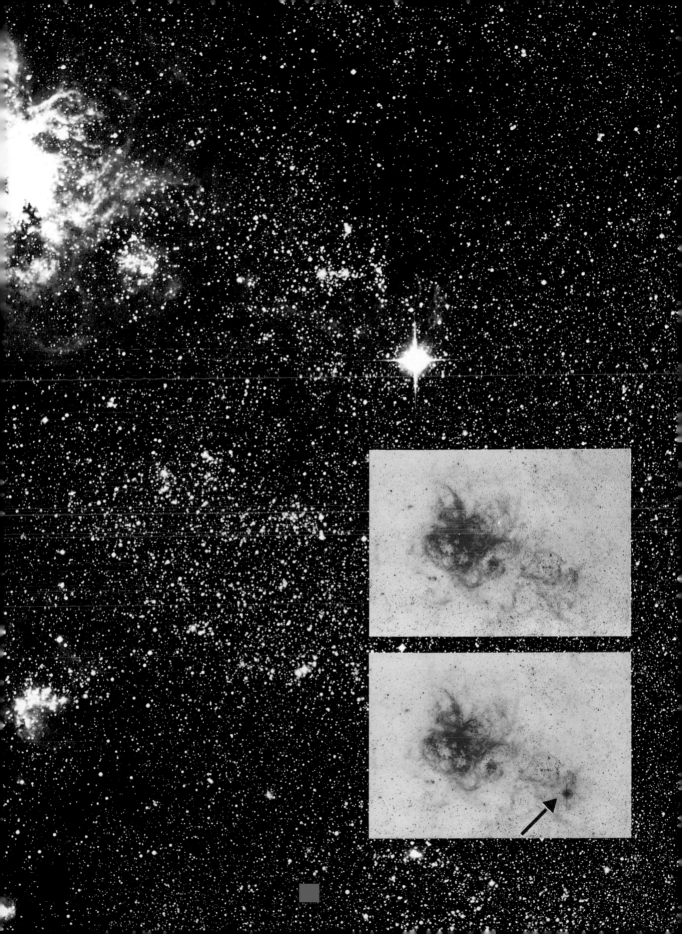

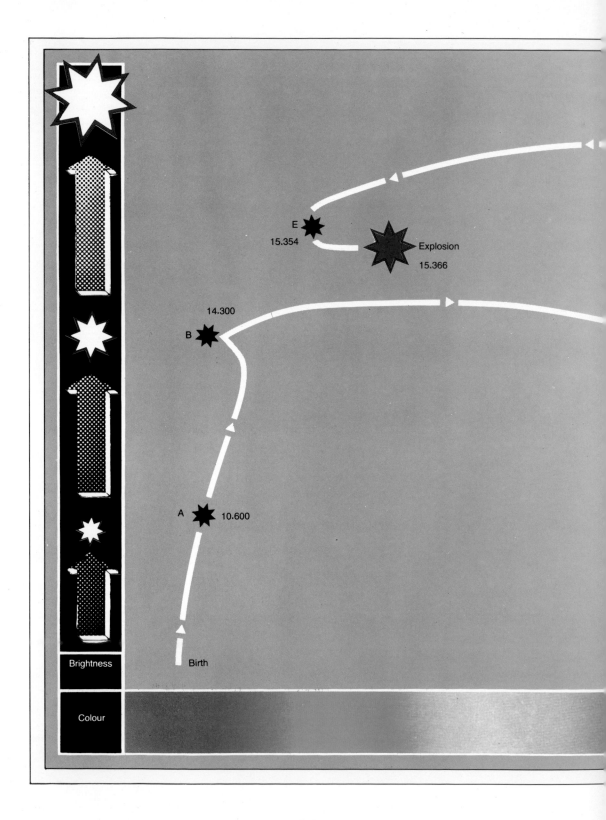

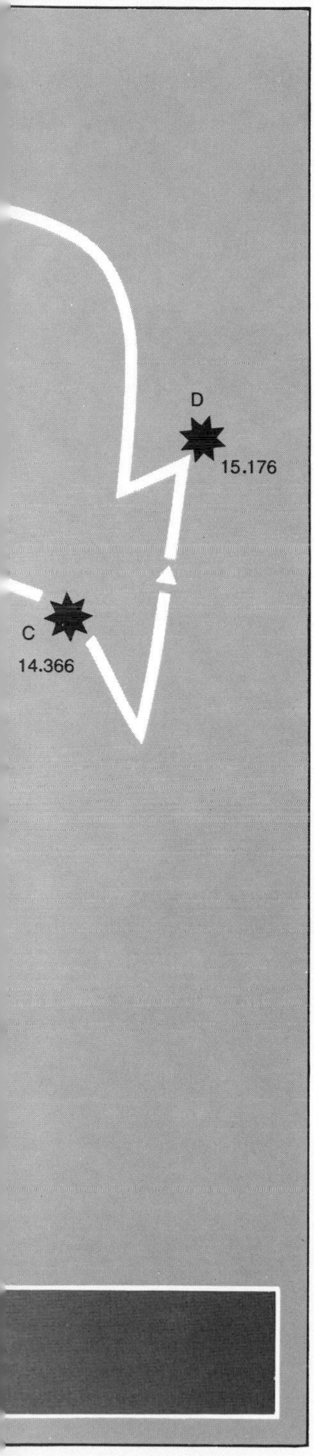

D
15.176

C
14.366

LIFE-CYCLE OF A SUPERNOVA

Theory says that young stars are blue and older stars are red. A high-mass red star will eventually exhaust its nuclear fuel, explode, and produce a supernova with a neutron star. But when astronomers looked at SN1987A a great surprise was in store, for the star which exploded in the Large Magellanic Cloud was blue, not red!

Based upon remarkable work by the astrophysicist S.E. Woosley and his colleagues at Santa Cruz, we can plot the life history of this unique star. It had a mass fifteen times that of the Sun, but a metal content of only one quarter. Over a period of nearly five million years the star altered in brightness and colour, and tracing these changes produces the track shown here. The age of the star in millions of years is given at points A, B, C, D, E and at the moment of explosion. It is interesting to see how a blue supergiant star can become a red supergiant (between points B and D) then turn blue once again before exploding. The star was only red for about 5 per cent of its life. The secret of how a blue pre-supernova came to explode lies in its relatively high mass yet low metal content.

A most interesting discovery with the world's largest optical telescope, on Mount Pastukhov in the Soviet Union, is the presence in Messier 33 of spectacular arc-shaped and bubble-like features of gas. A few of these are shown arrowed.

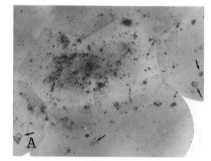

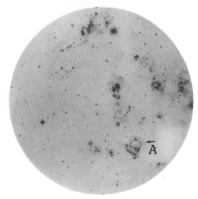

telescope in the southern hemisphere was trained on the Large Magellanic Cloud. Astronomer Ian Shelton had reported seeing a supernova! A star in the neighbourhood of the Tarantula Nebula had exploded sending gas rushing outwards with velocities of 17,000 to 18,000 kilometres per second. At such speeds, the blast wave would have covered the distance between the Earth and moon in a mere twenty seconds!

The Large Magellanic Cloud, with the more distant Small Magellanic Cloud, provide ideal opportunities for studying a complete galaxy. They contain young stars, old stars, dust clouds, glowing nebulae and star clusters – all the basic ingredients of our own galaxy. The Local Group of Galaxies also contains at least a dozen dwarf irregular galaxies. These, as the name suggests, are much smaller than the Magellanic Clouds. The remaining members of the group are elliptical galaxies, such as the two largest companions of the Andromeda spiral, together with dwarf elliptical systems.

Whenever I study galaxy photographs, one of my first questions is: What is man? And who am I in relation to these beautiful galaxies, and to the God who made them?

If there were no stars in the night sky what would be the challenge before us? I sincerely believe that the universe is as vast as it is, so that we might indeed search for its Creator.

The Small Magellanic Cloud is more distant and less bright than its larger brother, but it is easily visible to the naked eye from the southern hemisphere. It lies at a distance of some 200,000 light years.

But in the end the amazing truth is that the Creator has sought *us* out! The wonder 2,000 years ago was not the miracle star which hovered over Bethlehem, but the birth of Jesus in a rough stable below. In Christ God became man. The Creator came among us to share our humanity. The Son of God was born to serve us and to begin his restoring of spoiled creation.

'As we peek through giant telescopes into our huge galaxy and beyond into billions of other galaxies we become insignificant in the light of the vastness of the universe. But . . . as we look at Jesus we become mammoth in meaning. Jesus, the Creator, became a creation and died for us that we may be eternally significant.' Dr. Loren Cunningham.

We are significant because God thinks us of sufficient value to send his Son.

ISLANDS IN SPACE

THE MILKY WAY IS VAST ENOUGH, BUT IT IS ONLY ONE OF MANY GALAXIES IN THE UNIVERSE.

In the past the word 'nebula' was used to describe any fuzzy celestial object whether it was a star cluster, cosmic cradle or a galaxy. In Latin it simply means a cloud, and by the close of the eighteenth century the discovery and cataloguing of such patches of light in the sky had reached full swing. In 1755 the philosopher Immanuel Kant said that nebulae could be island universes or Milky Ways in their own right, but most astronomers rejected this idea. The

At the beginning of this century astronomers were still unsure whether 'nebulae' were objects within our Milky Way or were separate collections of stars many millions of light years away. The Whirlpool Galaxy shown here is now known to lie 35 million light years from us.

objects referred to by Kant are found in all directions in the sky except where interstellar dust obscures them. They should be distinguished from the gaseous nebulae – such as the Lagoon Nebula, the Trifid Nebula and the Veil Nebula – which *are* known to be part of the Milky Way.

The debate continued right up until the 1920s. One group believed that these objects all lay inside our own galaxy and that the Milky Way comprised the sum total of the universe. Others, such as Heber D. Curtis of the Lick Observatory and Knut Lundmark of Sweden, supported the island universe hypothesis. As David Bergamini recalls, 'Curtis undertook to announce to the world what seemed to him unequivocal evidence that nebulae containing faint novae are separate galaxies. But the world – at least of astronomy – was not yet ready to accept the huge universe that Curtis had to offer.'

In this historical document (1917), amended in his own handwriting, Curtis correctly argues that 'the spiral nebulae are, in effect, separate and very remote universes'.

It took the genius of American astronomer Edwin Hubble to resolve the controversy. He presented his exciting findings on certain variable stars (called Cepheids) in the Andromeda spiral to the American Astronomical Society on 30 December 1924. Cepheid variable stars are stars whose brightness varies over a period of days in a

Photographs of galaxies reveal great differences in the extent to which the central bulge dominates the overall appearance. Whereas the central bulge is exceptionally prominent in the Sombrero Hat Galaxy (left), it is less dominant in galaxies such as Messier 81 (centre) and is very small in Messier 101 (bottom).

Continues on page 104

*An example of an Sa
galaxy is the Sombrero
Hat Galaxy which has
a very large bulge and
tightly wound arms.*

*The bulge in the
Sombrero Hat Galaxy
is so conspicuous, and
the inclination so high,
that Dr David Malin
of the Anglo-Australian
Observatory has
employed a special
photographic technique
to 'remove' part of the
bulge light. This reveals
the otherwise almost
hidden and delicate
spiral arms which coil
around the bulge.*

Galaxy NGC 1300 does not fit into the simple scheme of Sa, Sb or Sc. About a third of spiral galaxies show bars running through their centres, and are called 'barred spiral galaxies'. After noting whether or not the galaxy has this central bar, the criteria for grouping such barred spirals into categories follow the same a, b, c notation as for unbarred, normal spirals.

regular pattern. They serve as excellent light beacons in space and by comparing their light output astronomers can calculate relative distances. Imagine looking at two light-houses, A and B, equipped with identical lamps. If light-house B is twice as far away as lighthouse A, the light received from B will be only one quarter of that from A because light obeys what is known as an inverse square law. Conversely, if we did not know the distance of light-house B, but observed its light output to be only one quarter that from lighthouse A, then we could deduce that it was twice as far away.

Hubble applied this argument to determine the distance of other galaxies. Cepheid stars in our Galaxy belong to a class of very luminous 'supergiant' pulsating stars. Hubble found a Cepheid variable in the Andromeda Spiral which was pulsating with the same period as a variable in the Small Magellanic Cloud. But it had only an eightieth of the brightness. Hubble concluded that, provided Cepheids of a given period have identical properties, the Andromeda Galaxy was nine times further away than the Small Magellanic Cloud – far beyond the confines of our Milky Way. The mystery of the distances to nearby galaxies had finally and decisively been unlocked. A new era in astronomy had begun.

In galaxies of Hubble type Sc the bulge is small while the arms follow a very open pattern.

Hubble grouped normal spiral galaxies into three categories, Sa, Sb and Sc, according to:

- the size of the central bulge compared to the overall diameter of the galaxy

- the tightness of the spiral arm pattern

- the visibility of stars in the spiral arms.

The Andromeda Spiral, Messier 81 and NGC 4622 are all Sb galaxies. The prominence of the central bulge is reduced and the arms are less tightly wound around the bulge. More stars can be seen in the spiral arms themselves. Bulges always consist of old stars while spiral arms – the locale of stellar birth – contain the young stars.

COLLISIONS IN THE SKIES

WHEN LOOKING AT A CLUSTER OF
GALAXIES, IT IS TEMPTING TO THINK
THAT THE SPACE BETWEEN THE
GALAXIES IS EMPTY. BUT THIS IS NOT
AT ALL TRUE.

'Rich' clusters of galaxies, that is, clusters consisting of
many galaxies, are generally X-ray sources. Measurements
of X-ray radiation from space show that a hot intergalactic

*Abell 1060, named after
the American
astronomer G.O. Abell,
is an example of a rich
cluster of galaxies
which is also an X-ray
source. The X-rays do
not arise from any
particular galaxy or
galaxies, but from the
whole central region of
the cluster.*

One of the spiral galaxies in the Abell 1060 cluster shows signs of stripping, which occurs as the galaxy ploughs its way through the hot gas pervading the cluster. The Abell 1060 cluster lies in the constellation of Hydra and is 150 million light years away.

gas spreads itself out in the cluster. This sometimes happens in a clumpy fashion while at other times the gas is spread out more evenly, with a concentration near the cluster centre. The space between the galaxies in a rich cluster is far from empty. In fact, the intergalactic gas can contain as much, if not more, mass than is in the galaxies themselves.

The intergalactic gas may be invisible to sight, but the temperature of the gas varies between 10 and 100 million degrees Centigrade! The presence of intergalactic gas will naturally affect the appearance of galaxies lying within such a cluster. Spiral galaxies, for example, often appear to be 'stripped' by the intergalactic gas and may look rather anaemic.

The nearest of the giant clusters of galaxies lies in the constellation of Virgo and it covers a region in the sky over six degrees in radius, or twelve times the diameter of the full moon. The Virgo cluster has been extensively studied in an effort to calculate the scale size of our universe. It is believed to lie some 60 million light years away and it also

△ The Virgo cluster, of which this photograph shows the central region, is the nearest of the giant clusters of galaxies. Note how the central spiral galaxy has been affected by the gravity of another very close galaxy.

shows how galaxies close to each other can be distorted in shape by the effects of gravity. Such gravitational encounters are relatively frequent in giant galaxy clusters.

Not only are there gravitational encounters within clusters of galaxies but also between 'isolated' galaxies. The two galaxies NGC 4038 and NGC 4039 lie relatively close together in space, and interacted with each other as a result of gravitational attraction. This produced a magnificent set of feelers or antennae reaching out five times the diameter of our Milky Way Galaxy!

A closer look at the gravitational encounter in Virgo. The spiral galaxy NGC 4435 has been distorted by the passage of its neighbouring elliptical galaxy NGC 4438. Material has been flung far from the centre of the parent spiral.

The feelers from the interaction of NGC 4038 and 4039 extend half a million light years from tip to tip, five times the diameter of our Milky Way Galaxy! This pair of galaxies is 50 million light years distant.

LOOKING AT THE PAST

HOW FAR DOES SPACE EXTEND? IS THERE AN END TO THE OBSERVABLE UNIVERSE?

The Fornax cluster is 100 million light years away and centres around a large elliptical galaxy, NGC 1399. A closer inspection shows that this galaxy is surrounded by a collection of tiny star-like images. (These should not be confused with the many elliptical galaxies which lie around NGC 1399 and appear on photographs as white blobs.) These tiny images are called globular clusters and each

Light from the Fornax cluster of galaxies has taken 100 million years to reach us. And remember that light can cover the distance between us and the moon in just over one second! The very conspicuous spiral galaxy (NGC 1365) lying to the right is also a member of the cluster.

The angular extent of NGC 6872 is so large that if it was a member of the nearby Virgo cluster (which is only 60 million light years away) it would cover an angle in our sky greater than does the full moon.

cluster contains between 10,000 and 100,000 stars. Our own galaxy is surrounded by a 'halo' of globular clusters.

Three times further away than the Fornax cluster lies another cluster, but this time in the southern constellation of Pavo. Here we see each galaxy as it was 300 million years ago! Several years ago I examined the angular size of the barred galaxy, NGC 6872, and to my amazement found it to be comparable to the angular size of nearby spiral galaxies. Yet spectral information clearly indicated it was 300 million light years away! The only conclusion was that NGC 6872 must be gigantic. I computed the length from the tip of the one spiral arm to the other. The answer: nearly 750,000 light years or seven times the diameter of our Milky Way!

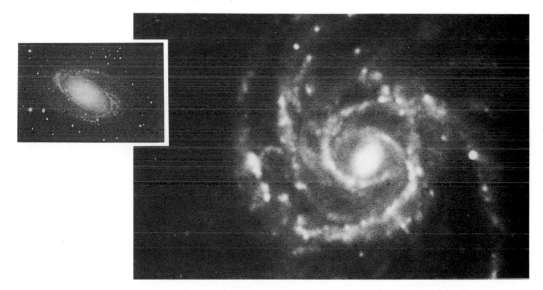

◁ *Galaxy NGC 1399, like our own galaxy, is surrounded by a 'halo' of globular clusters.*

Very large spiral galaxies are not uncommon. This photograph of one such spiral, NGC 309, also shows an inset of a galaxy we have already studied, Messier 81. But here Messier 81 is portrayed as if it were at the same distance as NGC 309. The difference in physical size is astonishing, especially bearing in mind that the diameter of Messier 81 itself is 78,000 light years!

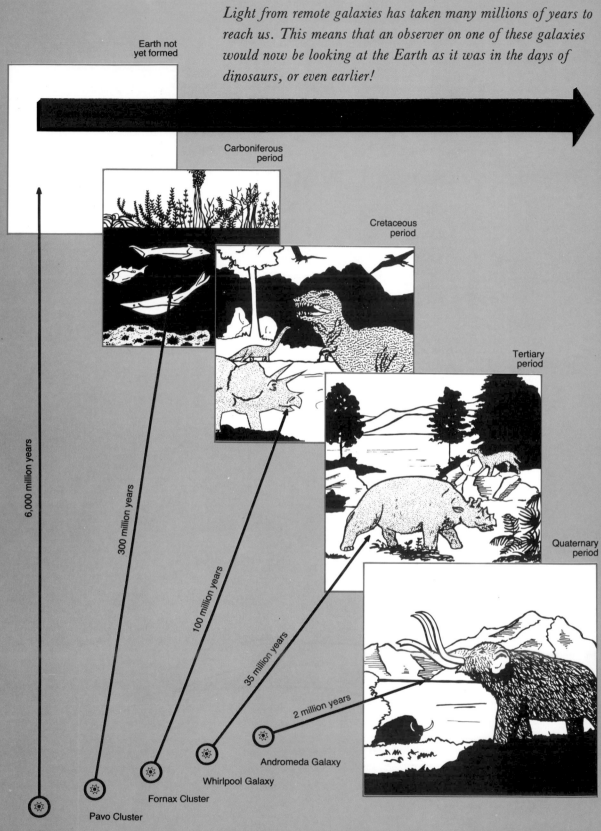

Light from remote galaxies has taken many millions of years to reach us. This means that an observer on one of these galaxies would now be looking at the Earth as it was in the days of dinosaurs, or even earlier!

Earth not yet formed

Carboniferous period

Cretaceous period

Tertiary period

Quaternary period

6,000 million years

300 million years

100 million years

35 million years

2 million years

Abell 370

Pavo Cluster

Fornax Cluster

Whirlpool Galaxy

Andromeda Galaxy

The extent of these arms is probably greater than in any other barred spiral known. A close inspection of photographs shows that this remarkable galaxy does have a companion. Were the very long arms produced by a gravitational encounter with its companion? It is well known that the motion of one galaxy past another can produce extended 'tail-like' features, such as in the antennae already discussed.

Since the publication of my results in 1979, a very large spiral has been studied by astronomer Vera Rubin at the Carnegie Institute of Washington. This galaxy has no companions, so gravitational encounters are not always needed to produce the vast size of these systems. Even in NGC 6872, the distinct kink seen only in one spiral arm could have developed after the entire galaxy itself had been formed.

The Abell 370 cluster is twenty times still further than the Pavo group. The galaxies are 6,000,000,000 light years away! At this distance, the amount of detail to be seen in each individual galaxy is far less than in nearby clusters, such as Virgo. In the light of so many galaxies, lying at such vast distances, we may feel humbled, even lost.

'When I look at the sky, which you have made,
at the moon and the stars, which you have
set in their places –
what is man, that you think of him;
mere man, that you care for him?

'Yet you made him inferior only to yourself;
you crowned him with glory and honour.
You appointed him ruler over everything
you made;

An amazing feature recently discovered in Abell 370 is a giant arc-like structure. This luminous 'rope in the sky' is large enough to wrap around the circumference of the Milky Way. It is especially puzzling how such a perfectly symmetrical arc can maintain its structure in the presence of so many cluster members, each moving within the cluster and having differing gravitational pulls on the arc.

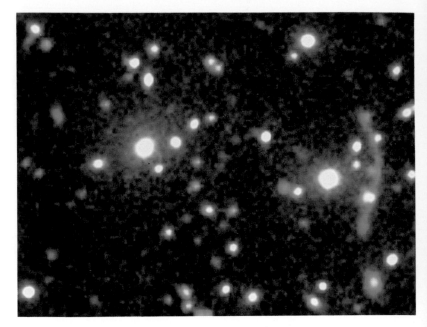

you placed him over all creation:
sheep and cattle, and
the wild animals too;
the birds and the fish and the creatures
in the seas.

'O Lord, our Lord, your greatness
is seen in all the world!'

The Eighth Psalm

The psalmist understands how we feel. We inhabit a tiny planet among myriads of far-flung galaxies. But the psalmist does not feel despair, only a deep sense of wonder. He knows that *God* created him in the same way as he created the infinite universe. However small and insignificant we may feel, we may also be certain of the value the Creator places upon us.

Extragalactic Radio Beacons

SOME GALAXIES GIVE OUT ABNORMALLY LARGE AMOUNTS OF RADIO ENERGY. THESE ARE KNOWN AS 'RADIO GALAXIES'.

Just as we can see galaxies through an optical telescope, so we can capture an image or 'picture' of the radio energy they emit by using a radio telescope. For some radio galaxies their visual and radio pictures are much the same, while others have unique properties that may be linked to their unusual radio luminosity. Radio galaxies are part of a

Radio telescopes are used to receive the radio energy emitted by galaxies. The telescope at the Hartebeeshoek Observatory is ten storeys high when pointing at the zenith.

The diameter of the dish is 26 metres – large enough to detect PKS 2000–330 15,000 million light years away.

class of galaxies called 'active galaxies'.

The nearest strong radio galaxy to us lies in the constellation of Centaurus, and is known as NGC 5128, or Centaurus A. An object of extraordinary astrophysical interest, Centaurus A is believed to be the outcome of a collision between a giant elliptical galaxy and a spiral galaxy. This would account for the curious optical appearance of the system, and was first suggested by two astronomers, Baade and Minkowski, in a paper published in the *Astrophysical Journal* in 1954.

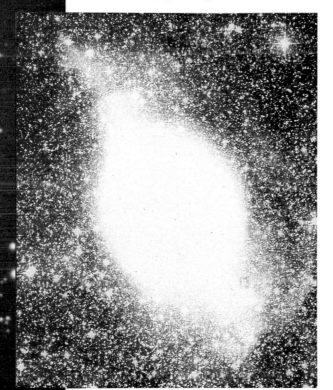

A 'deep' optical exposure of this curious system shows that even optically NGC 5128 is much larger than conventional exposures would reveal. The myriads of stars seen in this photograph belong to our own Milky Way Galaxy as we look at NGC 5128 some 10 million light years away.

Recent observations have shown that the elliptical component of NGC 5128 is encircled by a nearly complete ring of young blue stars. The stars are distributed in a disk located in the obscuring dark 'lane' running across the centre of the galaxy. The ages of these stars have been estimated to range from 10 million to 50 million years.

117

The movements of the gaseous material in the disk show that the disk was probably formed by the capture of a spiral galaxy by an elliptical one. But if this were so we should still be able to see the nucleus of the captured spiral. Observations of colliding galaxies show that galaxy nuclei are able to survive such encounters. Two candidates for the surviving nucleus have been suggested: a very faint galaxy called MGC 7-28-3, or a low surface brightness galaxy A1332-45. But both these have now been shown to lie at about twice the distance of NGC 5128 itself.

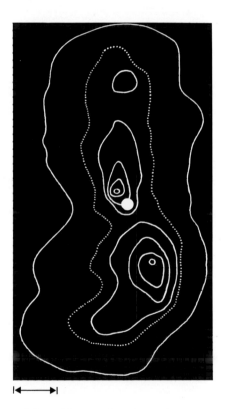

A radio 'picture' of Centaurus A. The visual part of the image is shown as a white central spot, with lobes of radio emissions extending on either side. The scale line represents 330,000 light years.

The elliptical galaxy is the centre of a great complex of radio-emitting lobes, which if visible to the naked eye, would extend ten degrees, or twenty times the diameter of the full moon in the sky.

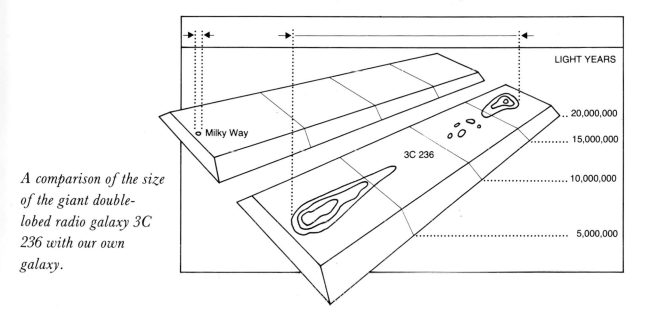

A comparison of the size of the giant double-lobed radio galaxy 3C 236 with our own galaxy.

The lobes of the radio galaxy 3C 236 extend nearly twenty million light years! A double-lobed structure is typical of many radio galaxies, which means that the radio waves are emitted mostly from two zones (or lobes) located far to either side of the object.

An amazingly large galaxy? Yes.

In an amazingly large universe? Yes.

Telling us of an amazingly all-powerful God.

We live in a world made by a personal God. A God for whom it is as easy, or as difficult, to make a galaxy as it is to make a flower, or a blade of grass.

THE BIG BANG

WE LIVE IN A DYNAMIC UNIVERSE. A
UNIVERSE WHICH IS CONSTANTLY
EXPANDING. GALAXIES, EXCEPT FOR
THE VERY CLOSEST ONES, RUSH AWAY
FROM US AT SPEEDS WHICH ARE IN
DIRECT PROPORTION TO THEIR
DISTANCE. THE FURTHER AWAY A
GALAXY, THE FASTER IT MOVES.

In 1929, Hubble published a famous paper in the *Proceedings of the National Academy of Sciences*. That paper, together with further observations published in 1931, established beyond any doubt that the universe is in a state of continual expansion. Galaxies recede at speeds dependent on their distances from us. One of the most distant objects ever photographed, 15,000 million light years away and almost at the very boundary of our observable universe, is PKS 2000–330. It moves away from us at an amazing 274,800 kilometres per second, which is over 90 per cent of the speed of light!

NGC 2997 recedes from us at 800 kilometres per second, and is 52 million light years away. The recession velocity of NGC 4622 is five times greater so this galaxy is five times farther away, at a distance of 260 million light years. These velocities can be measured with spectroscopic equipment attached to an optical telescope.

The most widely accepted model of our expanding universe is of a 'big bang'. Unfortunately this phrase suggests that all the matter of the universe once existed *somewhere* as a super-dense ball of matter which suddenly exploded, scattering debris thoughout space like an exploding bomb. But this picture is too simple. It presup-

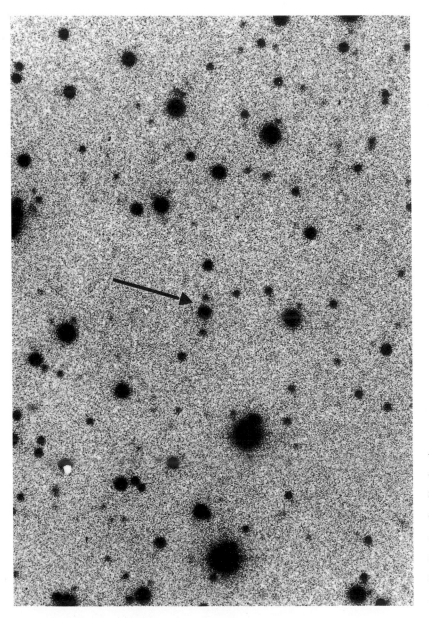

From PKS 2000–330 light has taken 15,000 million years to reach our eyes. We are indeed looking back to a time very close to when our universe came into being.

poses that space already existed before the explosion and that the big bang was merely the mechanism that flung the stars and galaxies out into space. It suggests that before the bang there was empty space, like an empty room, with a highly concentrated dot of matter at its centre. After the bang the galaxies moved out to fill space, like furniture fills a room.

But the big bang is also concerned with the formation of space itself. There was no 'room' within which the explosion could occur! Analogies are inadequate but it is something like representing the early universe as bread dough expanding as it is baked. But note that in this analogy the dough is not 'matter' and the oven in which it is baking 'space' for the dough represents *both* space and matter. Initially the dough is concentrated together, but as the bread rises the dough expands and becomes less dense. If you could run the baking process backwards then every part of the bread would contract. Yet at every stage the bread is bread; it is only the density at all points which varies.

So, in cosmology, the big bang represents a period of time in the early history of the universe when matter everywhere, at all points, was concentrated at immense density. Going backwards in time from today we would see the present outward-moving galaxies contracting, and at *all* points in the universe the density of matter would become higher and higher. We would only have to travel back in time, staying where we are, to find ourselves immersed in the highly dense big bang state. If the universe is infinite now, then it has always been infinite. If it is finite now, there is and was no 'beyond' in the usual sense of the word.

Twenty thousand million years ago the density at any one point was immensely high, and this is what cosmologists mean when they refer to 'the age of the universe'. At that period temperatures exceeded 10,000 million degrees!

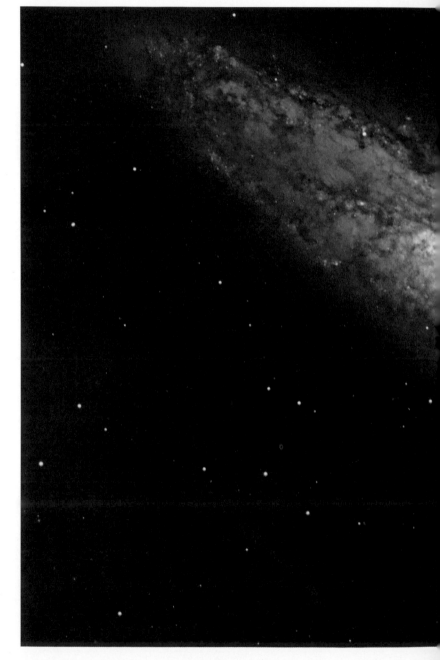

Remarkable work has shown that the visible boundaries of spiral galaxies do not, by any means, represent the total extent, or contain the total mass, of the spirals. Spirals of all Hubble types must be surrounded by vast haloes of dark matter. If they are not, they simply would not rotate in the manner which is observed!

A question I am often asked is: 'Will the universe continue to expand forever, or will the expansion eventually stop so that the universe starts collapsing in on itself again to another Big Bang?'

Calculations show that the density of *visible* matter in our universe, including all the stars, galaxies and clusters of galaxies, is not high enough for gravity to stop the universe from expanding forever.

But most astronomers believe that the greater part of the mass in the universe is 'dark', that is, invisible. The first hint that something was amiss came fifty years ago, when Fritz Zwicky noticed that galaxies in the Coma cluster are moving about much faster than expected. They are moving so fast that the galaxies should be flying apart! Unless, that is, there is sufficient dark matter to keep the cluster bound as a unit.

The most conservative possibility for the composition of the dark matter is that it is 'baryonic', or built up from the atomic particles (such as protons and neutrons) from which all astronomical objects are made. It is invisible because it is not radiating at any wavelength with enough intensity to be detected on Earth. Dark matter lurking in the form of black holes would also be baryonic. But there is a distinct possibility that the dark matter is 'nonbaryonic' and composed of exotic atomic particles such as neutrinos, axions or photinos.

Calculations based upon the density of visible mass favour a universe which has occurred only once. The scenario of a universe which expands, then collapses, expands, then collapses, and so on, is ruled out. But if 90 to 99 per cent of the universe is made up of dark matter, then the combined density of visible and dark matter could be sufficient for gravity to halt the expansion at some future stage. The existence of this dark matter is no longer disputed. But exactly how much there is remains unclear. How intriguing

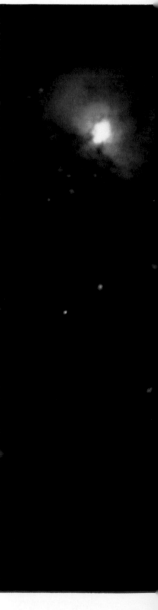

to think that the most abundant constituent of matter in our universe could be totally invisible!

In the initial formation of the universe an extremely delicate balance had to be established between the densities of the 'blobs' of matter destined to form galaxies, and the expansion rate of the universe. If the early universe ex-

The early universe expanded at just the critical rate to produce galaxies and planets.

panded too slowly, regions of higher density (which could have formed galaxies) would have had time to collapse in on themselves. Instead of galaxies they would have formed black holes. Today there would be no constellations, no suns and no Earth.

On the other hand, if the early universe expanded too quickly, the same regions of matter of higher density would have continued to expand outwards indefinitely. The star systems would not have been held together by gravity to form galaxies! The balance is critical. A reduction in the rate of expansion by only one million millionth when the temperature was 10,000 million degrees would have caused the universe to recollapse when its radius was only a tiny fraction of its present value.

Is our universe special? Could we still be reading this book if the basic physical forces in our universe were significantly altered?

The answer to the first question is yes, and to the second, no. We live in a finely tuned universe. By very reason of our existence, the universe must have special, and not arbitrary, properties.

For example, consider the implications of lowering the electromagnetic force. Electrons would no longer be bound to atoms, and we would have a universe where no chemical reactions were possible! On the other hand, if the electromagnetic force was significantly increased, electrons would lie trapped inside the nuclei of atoms, and again, no chemical reactions would take place.

If gravity was increased nuclear reactions in the cores of stars would be so rapid that the lifetimes of stars would be very short. If gravity was much weaker than at present

The more we learn about
the wonders of the Universe,
the more clearly we are going
to perceive the hand of God.

Frank Borman
Apollo 8

FORCES IN THE UNIVERSE

Four types of basic forces cement our universe together:

- gravitational forces

- weak interaction forces which operate on atomic particles over very short ranges

- electrical or electromagnetic forces

- nuclear forces which hold together protons in the atomic nucleus. Protons are positively charged and if unrestrained would move away from each other.

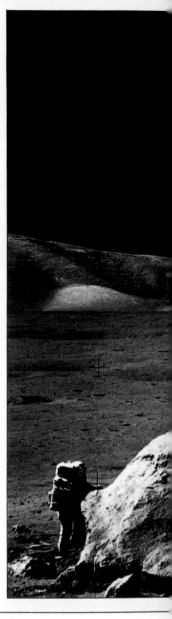

During December 1972 Apollo 17 touched down at Taurus-Littrow on the Moon. The commander Eugene Cernan took this photograph of his colleague Harrison Schmitt.

then stars would not get hot enough for nuclear reactions to start and we would have no suns!

For human beings to exist there must be limits to the diversity of physical laws and parameters that have governed the universe's development ever since its beginning.

Personally, I see this as the finger of God in the situation in a most special way. He created a perfect early universe expanding at just the critical rate to avoid recollapse. Was this so that human beings, the crown of his creation, could later live?

Continues on page 135

'There is a kind of religion in science; it is the religion
of a person who believes that every event in the universe
can be explained in a rational way as the product of some

previous event. This faith is violated by the discovery that
the world had a beginning under conditions in which the
known laws of physics are not valid. When that happens,

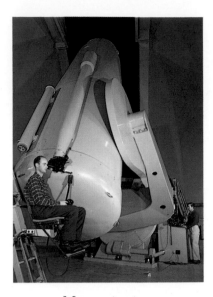

Many scientists acknowledge that their scientific studies cannot tell the whole story.

the scientist has lost control. He reacts by ignoring the implications, or by trivializing and calling it the big-bang, as if the universe were a firecracker.

'For the scientist who has lived by his faith in the power of reason, the story ends like a bad dream. He has scaled the mountains of ignorance; he is about to conquer the highest peak; and as he pulls himself over the final rock, he is greeted by a band of theologians who have been sitting there for centuries.'

Dr Robert Jastrow, former Director of NASA's Goddard Institute for Space Studies.

The Ghost of Copernicus

THIS UNIVERSE IS FINELY TUNED. HAS THIS TUNING ANYTHING TO DO WITH OUR HUMAN LIVES OR IS HUMANITY AN IRRELEVANT SIDESHOW?

It is often said that human beings do not occupy any special position in the universe. But the big bang model rests on two assumptions. One is a grand extrapolation of the fact that the Earth is not the centre of the universe. It is

The Earth is no longer believed to be the centre of the universe. But does this mean it is not special at all?

called the assumption or principle of 'homogeneity', which simply asserts that we do not live in any preferred position in the universe.

The reasoning behind this assumption goes back to Copernicus, who in the sixteenth century claimed that the Sun and planets were not moving around the Earth, as previously believed. Since then we have known that the Earth is not the focus of the solar system or the centre of the universe.

But can we justifiably extrapolate this to say that we do not live in *any* preferred region of the universe?

World-famous cosmologist George Ellis comments: 'While the historical roots of this assumption are well-known and based in valid experience, they do not reasonably justify extravagant extrapolations of the idea. In fact, it is manifestly invalid even in the solar system: our life could only have evolved on a planet situated within a very narrow range of distances from our Sun, and accordingly *does* occur in a very special situation within the solar system. *A fortiori*, life like ours could only have evolved at very special places in the galaxy.'

Ellis is saying that even within an evolutionary framework, we are not justified in assuming we could live equally well at all places in the universe. It is one thing for astronomers to recognise that our position in the universe is not central or special in *every* way, but quite another for them to

A camera lens was left open for several hours to record the movement of the stars. Each trail is, of course, the result of the Earth rotating about its axis and not the movement of the stars around the Earth.

assert that our locale is not special in *any* way. This claim is simply incorrect.

I have in my files a preprint of a paper written some years ago by a highly respected worker and his collabora-

The conventional big bang model of the universe is not the only possible universe model. Our universe contains galaxies and this is a serious problem in the standard big bang cosmology. Galaxies, and the nebulae inside them, should not have formed!

tors on a new model of the universe. The paper was entitled: 'The Universe Probably is Expanding. (But maybe we're near its centre)'. All papers published in leading science journals go through an international refer-

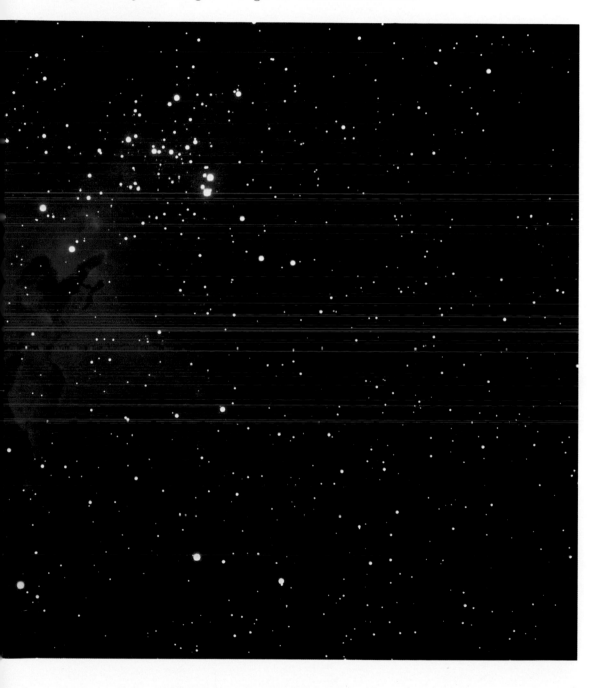

eeing system, and the title must have been too controversial. It was changed to read, 'The Expansion of the Universe.' The rest of the paper was published unchanged. Believing in a preferred position would, to many, be threatening.

But can we ever say the universe is centred on our presence? Consciously or subconsciously, thoughts about a Creator and the creation enter in. Science always claims to be completely objective, but on this issue I sometimes find objectivity curiously lacking. For if we strictly follow the Copernican Principle (that is, that there are no 'preferred' regions in our universe) it follows that the density of matter of any one region should be exactly the same as that of any other region. Galaxies, which are areas of increased density compared to their surroundings, should never have formed!

A great deal of effort has gone into tracing the origins of galaxies back to minute fluctuations in density in the very first moments of the universe. One recent idea is the 'inflationary universe'. This model is identical to the conventional big bang except that at very early times the universe 'inflated' or 'ballooned outward' much faster. In this scenario, centres around which matter could condense to form galaxies survived the inflationary phase. Such models do not insist on a homogeneous distribution of mass at the start.

The existence of nebulae and galaxies today implies that the universe could not always have been homogeneous. There must have been pockets of higher density material.

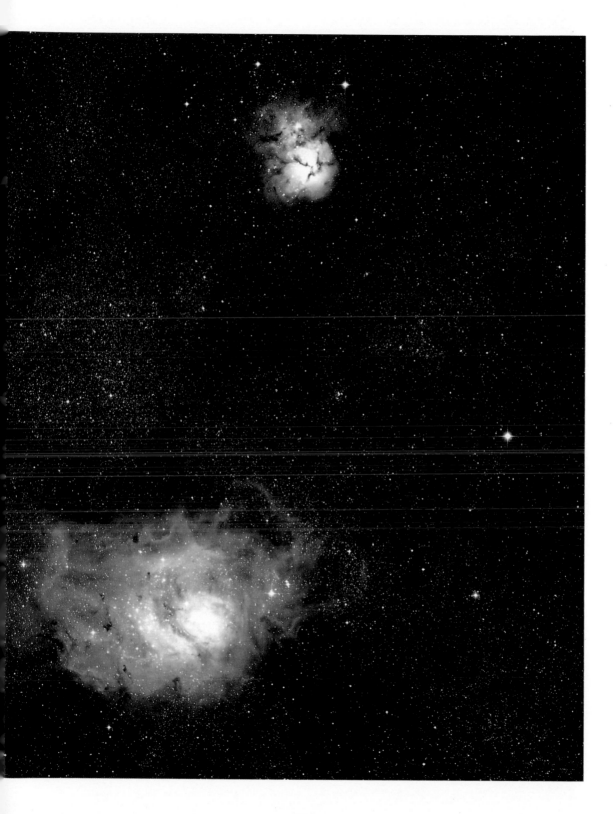

The simplest understanding of our universe is of uniform expansion following an immensely dense big bang state. This still remains the most popular model, and is the one we shall continue to discuss here. But it is important to remember that it is not the only working model. We may be living near the centre of some inhomogeneous but expanding universe, a universe which differs radically from the big bang.

A growing number of prominent astronomers, cosmologists, and physicists are challenging the Copernican perspective of the universe. They have reintroduced into cosmology an 'anthropic perspective' that views certain characteristics of our universe as strongly linked with our own human existence. We have already seen that the early universe expanded at just the critical rate to harbour mankind billions of years later.

Two such cosmologists are John Barrow and Frank Tipler, who refer to one version of the 'anthropic principle' as saying that, 'the Universe must have those properties which allow life to develop within it at some stage in its history'. Barrow and Tipler explain that, of all the possible mathematical models of the universe, the only realistic ones are those universes which could harbour life at some time in their development. On these grounds, an early universe expanding too fast, or too slowly, would be inadmissible.

In the Great Nebula in Orion a cloud of gas surrounds several very hot stars deep within the nebula. If the initial expansion rate of the universe had been lower these spectacular stars would all have been part of a large black hole.

145

Looking at the night sky, some might wish to echo the words of Bernard de Fontenelle (1657–1757): 'Behold a universe so immense that I am lost in it. I no longer know where I am. I am just nothing at all. Our world is terrifying in its insignificance.'

But could our universe harbour intelligent life if it were not as big, and as old, as astronomers find it to be?

The key point here is that the observable extent of the universe expanding from a highly dense big bang state is inextricably bound up with its age. Each year, our observable horizon increases by one light year, so that the created universe we find ourselves in must be at least 15,000 million light years in size because it is at least 15,000 million years old. Barrow and Tipler explain:

'The requirement that enough time pass for cosmic expansion to cool off sufficiently after the Big Bang to allow the existence of carbon ensures that the observable Universe must be relatively old and so, because the boundary of the observable universe expands at the speed of light, very large. No one should be surprised to find the Universe to be as large as it is. We could not exist in one that was significantly smaller.'

If it does take 15,000 million years after a hot big bang state for temperatures to be supportive of life, and for the Earth to be habitable, then the universe *must* be 15,000 million light years in extent, for it has been expanding for

'There exists one possible Universe specifically designed with the goal of generating and sustaining observers.' Barrow and Tipler on the Anthropic Principle.

146

Discovering The Creator

'As a matter of history modern cosmologists have been unable to think about the universe without thinking about creation. Postulating a universe seems to me to be almost the same as postulating a *Creator*.

'As we have seen repeatedly, we cannot formulate any science without reference to the *observer* and, again as a matter of history, progress in fundamental science has been made by increasingly recognizing the role of the observer. It seems to me therefore that we cannot think about the universe without the concept of *personality*. Cosmology requires, I venture to assert, the concepts of Creator and of personality, and together these mean God.

'There can be no cosmology without physical experience, and there can be no religion without religious

15,000 million years at the speed of light! The argument that the universe must be teeming with other civilizations than ourselves, simply because it is so immense, loses its persuasiveness: the universe has to be as large as it is just to support life on Earth!

The Anthropic Principle, so named by cosmologist Brandon Carter, is a 'reaction against exaggerated subservience to the Copernican Principle'. It takes the wind out of the sails of the outlook expressed by de Fontenelle. The principle recognizes that we live in a very finely tuned universe, where size, age and all the natural forces are not arbitrary, but are adjusted to support life. If it were not so, experience. Men exert themselves to the utmost to gain new experience of the physical world; likewise religious experience has to be sought.

'Any such experience that any of us may claim speaks to us surely of *purpose*. Whether it be the glory of a morning in springtime, or the beauty of a human face, or the vastness of the universe producing the specks that are ourselves, can we possibly believe there is no purpose in it?

'Some may consent thus far and be therewith content. But to others of us purpose is inseparable from person, and the Person of the Creator is revealed in the person of Christ.'

Extracts Sir William McCrea's chapter in *Cosmology, History and Theology*, W. Yourgrau and A.D. Breck (eds.), Plenum Press, 1977.

we would not be here.

Mankind could not live in an arbitrarily large or small universe. Our universe is a home. Designed, I believe, by the hand of God.

NAMING THE STARS

Astronomical objects are referred to by their place in a number of recognized lists or catalogues. Messier 8 (or M8 for brevity) refers to the eighth object listed in the famous catalogue compiled in 1781 by the French astronomer Messier. The following catalogues are referred to in this book:

M *Messier's Catalogue*, 1781

NGC J.L.E. Dreyer's *New General Catalogue*, 1888

HD *Henry Draper Catalogue* compiled by Annie Cannon, 1918–24

HDE *Henry Draper Extension* containing information on an additional 46,850 stars

MGC *Morphologicheskii Katalog Galaktik* by Soviet Astronomer B.A. Vorontsov-Velyaminov, 1962-74

3C *Third Cambridge Catalogue of Radio Sources*

PKS Radio sources which were first detected or discovered at the Parkes Radio Astronomy Observatory in Australia

A 'Anonymous' listing used when an object is not found in standard references such as the *New General Catalogue*

CONSTELLATION GUIDE

Description	Constellation
The Pleiades	Taurus
The Rosette Nebula	Monoceros
The Eagle Nebula	Serpens
The Lagoon Nebula	Sagittarius
The Trifid Nebula	Sagittarius
The Cone Nebula	Monoceros
The Orion Nebula	Orion
The Horsehead Nebula	Orion
The Crab Nebula	Taurus
The Veil Nebula	Cygnus
The Helix Planetary Nebula	Aquarius
The Ring Nebula	Lyra
Black hole candidate Cygnus X-1	Cygnus
Andromeda Spiral	Andromeda
Sombrero Galaxy	Virgo
Spiral NGC 4622	Centaurus
Sb Spiral M 81	Ursa Major
Triangulum Galaxy	Triangulum
Spiral NGC 253	Sculptor
Spiral NGC 2997	Antlia
Galaxy Messier 83	Hydra
Galaxy Messier 101	Ursa Major
Whirlpool Spiral	Canes Venatici
Barred Spiral NGC 1300	Eridanus
Large Magellanic Cloud	Dorado/Mensa
Small Magellanic Cloud	Tucana
Supernova 1987A	Dorado
The 'Antennae'	Corvus
Centaurus A	Centaurus
Interacting galaxies in Virgo Cluster	Virgo
Abell 1060 Cluster	Hydra
Pavo Group	Pavo
The Fornax Cluster	Fornax/Eridanus
Abell 370 Cluster with giant arc	Cetus

To determine where the stars and galaxies mentioned in this book may be found in the night sky, first look up the constellation using the table. Then find the position of the constellation with the star charts shown on the following pages. There are charts for both northern and southern hemispheres. Some constellations will be below the horizon for observers in high latitudes.

STAR CHART: NORTHERN HEMISPHERE

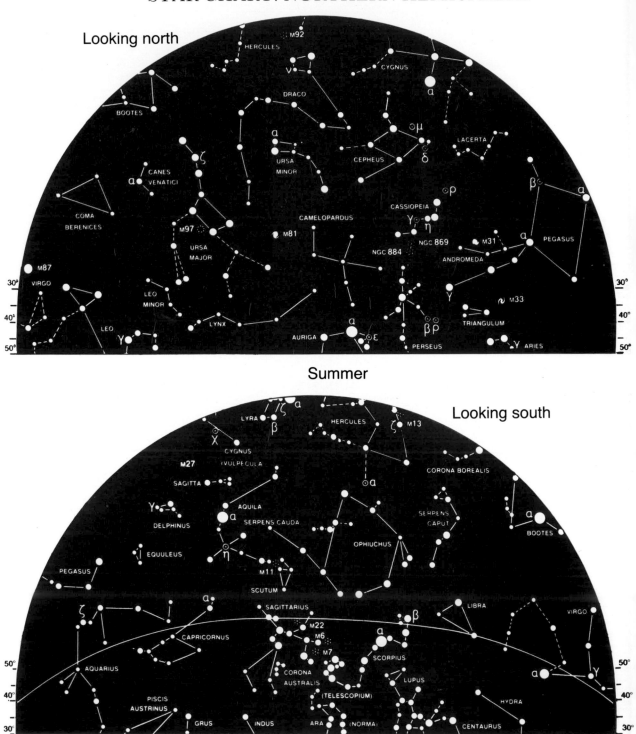

Looking north

Summer

Looking south

STAR CHART: NORTHERN HEMISPHERE

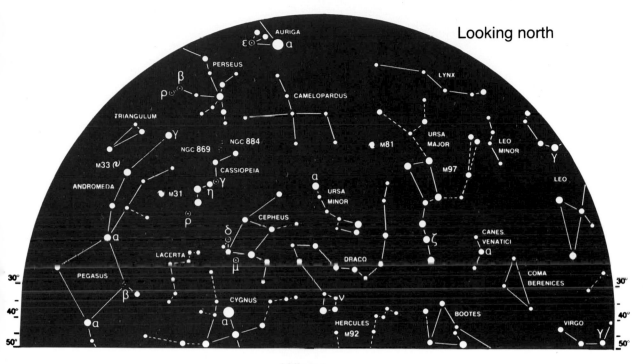

Looking north

Winter

Looking south

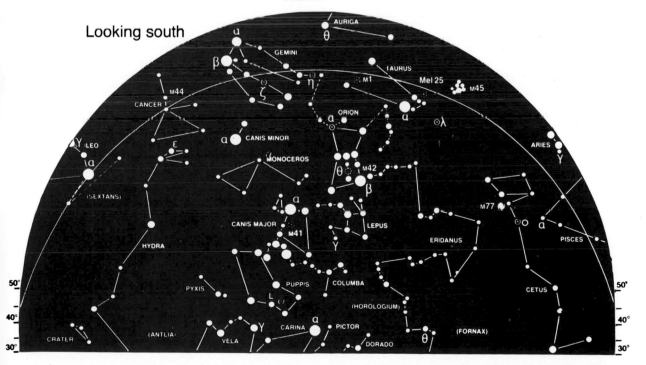

STAR CHART: SOUTHERN HEMISPHERE

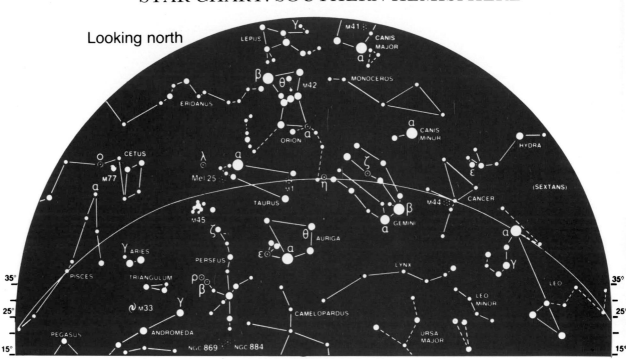

Looking north

Summer

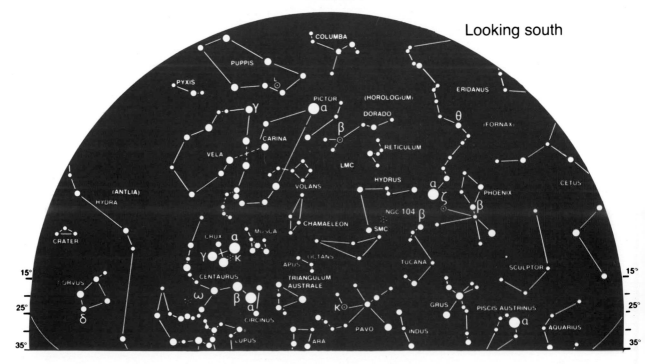

Looking south

STAR CHART: SOUTHERN HEMISPHERE

Looking north

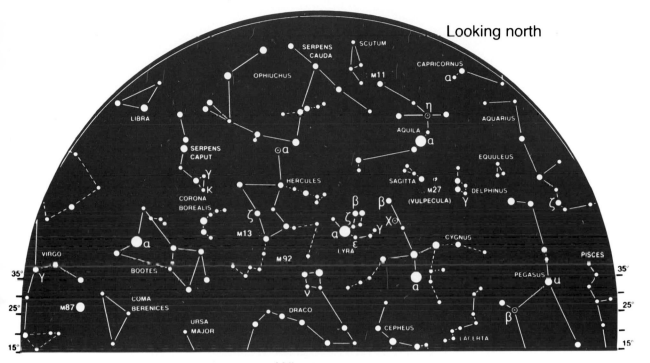

Winter

Looking south

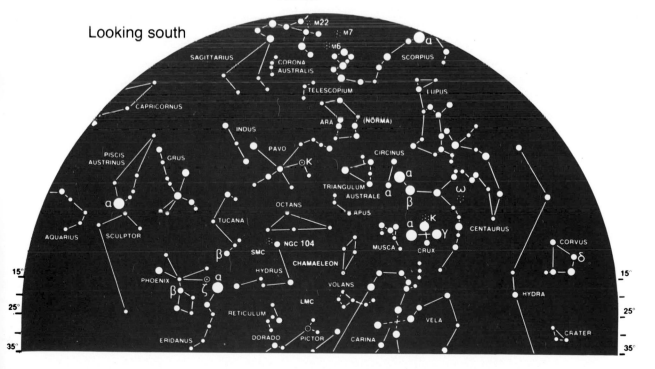

ACKNOWLEDGMENTS

To all those who have supported me in this project, both financially and in prayer, my deepest thanks.

Special thanks are due to Dr Alan Stockton, Staff Astronomer at the Institute of Astronomy, University of Hawaii, for reading through the entire first draft of this book, and for suggesting many helpful improvements. Sir William McCrea, Emeritus Professor of Astronomy at the University of Sussex, is warmly thanked for offering helpful comments on the cosmological content of this book. The first draft was also read by Dr Loren Cunningham and his encouragement and expert advice is much appreciated. I am most indebted to Mr L.C. Pouroulis for donating word processing equipment.

It is a pleasure to thank the many institutes and observatories for their sterling co-operation in supplying the colour transparencies and prints which are acknowledged below. Dr David Malin, Staff Astronomer at the Anglo-Australian Observatory, is especially thanked for generously supplying his exquisite colour work.

Finally, I thank my wife Liz, Irene, Faith, Jo, Anna, Veronique, Linda and Fido for all their assistance and hard work.

PHOTOGRAPHS

Anglo-Australian Observatory
6, 30, 40, 51, 54, 61, 62, 72, 81, 101 (top),
103 (bottom), 105 (bottom), 106, 107, 116, 121, 122, 125, 138

F. Apps
12

Astronomical Institute, Bratislava
38

BBC Hulton Picture Library
87

California Institute of Technology
39, 45, 48, 60, 76, 78, 82, 91, 104, 136

D. Block
115 (both)

L. Cohen, Griffith Observatory
111 (inset)

B.F.C. Cooper, R.M. Price and D.J. Dole
118

G. Courtès, H. Petit and J.P. Sivan, Observatoire de Marseille and
Laboratoire d'Astronomie Spatiale de Marseille
96

European Southern Observatory
58, 88, 89, 93 (colour), 102, 111 (top)

I.S. Glass, South African Astronomical Observatory
90

Hale Observatories
101 (centre), 101 (bottom)

S. Halliday and L. Lushington
37

Jet Propulsion Laboratory, Pasadena
11, 14 (both), 18, 20, 21, 22, 23, 25, 26, 27

Franz X. Kohlhauf
147

J. Kristian, Mount Wilson and Las Campanas Observatories
84

Lick Observatory, Santa Cruz
47, 74, 100

Life Picture Service
41

Mitchell Beazley Limited
152–155

Monumenti Musei E Gallerie Pontificie
132

National Optical Astronomy Observatories
42, 71, 77, 99, 114, 140, 141, 145

National Aeronautics and Space Administration
129, 131

Naval Research Laboratory, Washington
67, 68

Royal Astronomical Society
69

REFERENCES

J.D. Barrow and F.J. Tipler, *The Anthropic Cosmological Principle*
Clarendon Press, 1986
146

D. Bergamini, *The Universe*
Time-Life International, 1964
100

Loren Cunningham, private correspondence
98

G.F.R. Ellis in the journal *General Relativity and Gravitation*
Plenum Press, 1979
139

T. Ferris, *Galaxies*
Stewart, Tabori and Chang Publishers, 1982
50

R. Jastrow, *Have Astronomers Found God?*
New York Times Magazine, 1978 and The Reader's Digest, 1980
134–136